Palgrave Studies in Impact Finance

Series Editor

Mario La Torre, Department of Management,
Sapienza University of Rome, Rome, Italy

The *Palgrave Studies in Impact Finance* series provides a valuable scientific 'hub' for researchers, professionals and policy makers involved in Impact finance and related topics. It includes studies in the social, political, environmental and ethical impact of finance, exploring all aspects of impact finance and socially responsible investment, including policy issues, financial instruments, markets and clients, standards, regulations and financial management, with a particular focus on impact investments and microfinance.

Titles feature the most recent empirical analysis with a theoretical approach, including up to date and innovative studies that cover issues which impact finance and society globally.

Karamo NM Sonko · Mariama Sonko

Demystifying Environmental, Social and Governance (ESG)

Charting the ESG Course in Africa

palgrave
macmillan

Karamo NM Sonko
Heeno International
Banjul, The Gambia

Mariama Sonko
Jula Consultancy FZE
Sharjah, United Arab Emirates

ISSN 2662-5105 ISSN 2662-5113 (electronic)
Palgrave Studies in Impact Finance
ISBN 978-3-031-35866-1 ISBN 978-3-031-35867-8 (eBook)
https://doi.org/10.1007/978-3-031-35867-8

Cover credit: Scharfsinn/Alamy Stock Photo

This Palgrave Macmillan imprint is published by the registered company Springer Nature Switzerland AG
The registered company address is: Gewerbestrasse 11, 6330 Cham, Switzerland

To Binta (wife and mother), Nyarika Kemo (son and brother) and Amadu (son and brother) for their excellence in all they are as our family!

Foreword

Understanding the importance of ESG factors in Africa is to appreciate the complex and painful history of the continent and contemporary challenges. Therefore, this calls for the need to address the intersectionality of a range of influences and variables and, in South Africa, in particular, the enduring legacy of inequality to create sustainable, equitable, and inclusive economies.

At the time of publishing this book, Africa and the world are experiencing recovery from the unprecedented social and economic impacts caused by the COVID-19 (coronavirus) pandemic, coupled with ongoing rising global temperatures and the negative effects of climate change. With every section of humanity facing increasing crises regarding the existence of the human species and life as we know it, it is crucial for us to approach our responses to pandemics, natural disasters, and economic crises from a holistic perspective.

Preparing for the future urges us all to understand our individual duties to reinforce and galvanize collective action; that is our urgent and shared responsibility to play our part in determining environmental, economic, social, and political prosperity within a framework of sustainable development goals, as argued in this book.

ESG disclosures and ESG performance are now core to the ways responsible organizations operate as stewards of not only the economy but as shared custodians of our natural environment, local communities, good governance practices, and respectful labour practices.

As the youth develop a growing awareness of ESG consciousness, they will also start to believe more in 'putting money where your values are'. Investing in sustainable development becomes increasingly a key driver of innovation and new opportunities to create long-term value for African societies and the private sector will emerge. These principle-based initiatives are critical to advance the United Nations' Sustainable Development Goals (SDGs). A coalition between the private sector, public sector, and civil society through ESG disclosures and ESG performance offers a blueprint for peace and prosperity for all people and for our planet.

A considered partnership of companies implementing ESG and other organizations pursuing SDGs strengthens the urgency for African and all countries to recognize that ending poverty, inequality, and other deprivations, go hand-in-hand with intersecting and integrated strategies that improve peace, health, education, hunger, justice, and economic growth as well as climate change and the preservation of our natural heritage. This roadmap for prosperity and progress is meant to leave no one behind, as pointed out in this book. For this reason, how we better manage the world's resources for the good of all humanity in the years to come, is the next piece in the puzzle of human development.

However, when embracing development objectives, it is important to recognize that each country is different and therefore has different needs, challenges, and capabilities. The needs of developed and developing countries, established, and emerging economies are very different. Countries with citizens in extreme poverty have contrasting challenges to first world nations, including where to invest precious resources and brainpower, and focus policy and social mobilization efforts. These are particular to each State. Therefore, African countries should not simply copy and paste ESG measures to address their needs, instead they should be able to assess their circumstances, economies, and societies to determine what they should consider material to their urgent challenges, whilst also recognizing the global threats which they share with the rest of the world.

This study is a valuable contribution to understanding Africa's ESG challenges and at the same time a valuable starting point to understanding South Africa's unique situation, challenges, and opportunities from a range of contrasting perspectives. The book highlights the positive potentials for incorporating ESG as an essential element in organizational best practices and provides a beneficial tool to realizing a sustainable, equitable, and inclusive Africa and the world for all.

His Excellency Kgalema Motlanthe

Third President of South Africa

Preface

There has never been a better time in recorded human history to discuss issues of the environment, society (particularly health), and corporate governance than now, largely because of the COVID-19 pandemic. These issues are not new but COVID-19 has increased their urgency dramatically. Although the reported number of COVID-19 cases in Africa is lower than those of other continents, the emergence of the Delta, Omicron, and other new variants have pushed the African numbers up and have created even more serious causes for concern. This sad reality draws attention, with a greater urgency than ever before, to the roles of governments, businesses, civil society, and academia in preparing countries and communities to combat the pandemic as well as other challenges in order to create a better world for all. What we refer to in this book as *formal comprehensive* or holistic ESG has emerged in recent years as a response to concerns about the future of our world before COVID-19 and shall continue to increase in importance even after the pandemic has ended, as the clarion call for sustainability increases worldwide, including Africa.

The aim of this book is to discuss the ESG agenda in Africa. The continent has a very long way to go before fully understanding, adopting, and implementing ESG disclosures or reporting.[1]This is in spite of the huge

[1] Disclosure and reporting are used interchangeably in this book, but disclosure is also used as a noun.

challenges that it has been facing before the COVID-19 pandemic, it is facing during the pandemic, and shall continue to face after the pandemic.

The study is divided into four parts and eight chapters. It consists of an introduction to ESG, a strategic management approach to it, overview and relevance of it in Africa, importance and priorities during the COVID-19 pandemic, designing and reporting of ESG disclosures (frameworks, processes, standards, formats, issuers, investors, regulators, etc.), regional case studies comprising four countries, including an empirical analysis of ESG in South Africa, and finally the way forward for the key stakeholders—governments, investors, managers, workforce, NGOs and the communities. The findings on South Africa (Africa's leading ESG country by far) present significant propositions for ESG and sustainability-oriented practitioners and stakeholders in Africa and beyond.

Our study is distinctive because of several reasons. The authors of this book have more than two decades of experience in issues that comprise ESG today. These are environmental oversight, corporate social responsibility (CSR), corporate governance, community engagement, and government relations. Unlike most of the current discourse on ESG, this book is neither populistic/political, nor appealing only to sentiments or ethics profit-seeking corporations to act responsibly. Instead, it argues that ESG is largely in the interest of the companies and organizations themselves, in addition to benefitting the larger society and world in which they exist. Attacks (some extremely fierce) on financial capital in Africa come from numerous directions, but rarely does anyone tell or prove to the companies that "ESG is in your own interest because it can enable you to make sustainable business, whilst obviously enhancing growth in the national economy in which you exist". Not enough attention is also focused on the role of the government in encouraging and facilitating the execution of ESG disclosures in Africa. In short, in this study we view the survival of businesses, the wider economy, society, and world as intertwined. Furthermore, there is a need to look at how a pact can be formed in achieving what is being touted as shareholder capitalism, which is now being pushed forward as the international business concept for promoting ESG measures. However, we go beyond this "new" concept of stakeholder capitalism by putting the role of the state in Africa under the radar too.

Although research and publications have been done on many aspects of ESG over the years, the topic of *formal comprehensive or holistic* ESG is new and remains generally uncovered, especially in Africa. Therefore, this book is most likely to be the first ***scholarly researched book*** on the topic of *formal comprehensive or holistic* ESG in at least Africa, if not far beyond.

Our biggest challenge in writing this book was how to take into account the largely different comments of six reviewers on what to focus and how to focus on what they consider most important on a topic that is new in both the academic and practical worlds. In particular, there are those who believe that a book for practitioners is what is needed, whilst others take an academic orientation. We decided to harmonize both in order to attract a wider audience. Therefore, as can be seen from the contents of the book, we have combined the two by integrating multidisciplinary, theoretical, and rigorous empirical analyses with explanations and examples of how to practically design ESG reporting frameworks and understand related concepts. In addition, to regional case studies, we have carried out a survey, in the Gambia through the Chamber of Commerce and Industry (GCCI), which covers numerous companies in different sectors to determine the extent to which the actual knowledge and practice of ESG exist on the ground, in at least one African country without a stock exchange as an example. This is useful as the evidence in Chapter 6 suggests that the African knowledge about ESG/sustainability is highly skewed in favour of listed companies in countries with vibrant stock markets, because of the influence of global trends through these exchanges.

The authors wish to acknowledge the support of giants, dear friends, and dear colleagues who stood by us in this work, through encouraging words and invaluable comments. We wish to express gratitude to former classmates (Karamo's): Terry Fry, Executive Vice President for Global Strategy, Cadmus Group (San Francisco, CA); and Professor Ha-Joon Chang formerly of the University of Cambridge, now of the School of Oriental and African Studies (London). Thanks are also due to: Professor Haider A. Khan (our Professor!) of the Josef Korbel School of International Studies (JKSIS), University of Denver, CO; His Excellency Dr. Omar Alieu Touray, formerly of the Islamic Development Bank (Jeddah, Saudi Arabia) (now President of the Commission of the Economic Community of West African States (ECOWAS) (Abuja, Nigeria); Natasha Leger of ITF Advisors (Colorado Springs, CO); Pa Modou Baldeh, Ministry of Finance (Banjul, The Gambia); and Dr. Samsurin Welch, Lecturer, Associate—Circular Economy Centre, Judge Business School, University of Cambridge. The authors are very grateful to the supervisor of Mariama's M.Sc. dissertation research at Heriot-Watt University (Dubai Campus), Professor Faizul Haque, for his motivational guidance in her earlier South African empirical study at HWU. Thanks to Dr. Janvier Nkurunsiza of the United Nations Conference on Trade and Development (UNCTAD) (Geneva, Switzerland) for his invaluable comments on Chapter 7, particularly on the regression methodology and results. We are also grateful to Dr. Frank Boateng, Senior Lecturer and Head of the Department of Management Studies and Vice Dean of the Office of Research, Innovation and Consultancy, University of Mines and Technology (Tarkwa, Ghana). He and his students assisted us in collecting part of the information for all the country case studies in Chapter 6. Finally, we wish to thank all other colleagues and friends in different countries and institutions who commented on this work and contributed to it in different ways over a period of about three years whilst we were writing it.

Sharjah, United Arab Emirates

Karamo NM Sonko
Mariama Sonko

Contents

List of Figures

List of Tables

PART I

Global Overview and Relevance in Africa

CHAPTER 1

ESG: Global Importance, Origins, and Emergence

Abstract This chapter introduces ESG, defines it, outlines its evolution, and clarifies misunderstandings with regard to the topic. It reviews the literature, caused for its rapid rise and carries out an extensive survey of related concepts and activities in different parts of the world. The complex web around the topic is unravelled and the reader is taken on the journey of sustainability which leads to Africa through the rest of the book.

Keywords ESG · United Nations · World Bank · London Stock Exchange

> *Environmental, Social and Governance issues should be a priority for Boards and management. The advantages of proactively tackling ESG issues are significant. A robust ESG program can open up access to large pools of capital, build a stronger corporate brand and promote sustainable long-term growth benefitting companies and investors. There was a time when a public stance on ESG issues was a public relations tactic. That's no longer the case.*[1]

[1] "How Your Company Should Embrace Environmental, Social and Governance Issues," *Forbes*, November 21, 2018.

K. N. M. Sonko and M. Sonko, *Demystifying Environmental, Social and Governance (ESG)*, Palgrave Studies in Impact Finance,
https://doi.org/10.1007/978-3-031-35867-8_1

Environmental, social, and governance exchange-traded funds (ESG ETFs) alone topped $52 billion in November 2019, with 54 per cent of globally surveyed investors planning to increase their allocations of these assets in 2020.[2] *These aren't the only sustainable finance instruments seeing renewed interest: Green bonds, created in 2007 to fund projects with positive environmental benefits, have grown significantly. In 2019, the total value of green bonds issued grew to almost $255 billion, a nearly 50 per cent increase from 2018.*[3]

1.1 Introduction

There are not many social science phenomena or concepts that are as big and popular as Environmental, Social, and Governance (ESG) (Fig. 1.1), have spread as fast as ESG, and at the same time remain in a state of near-mystery or confusion equal to that of ESG. Despite its growing attraction, there is not much understanding about what this topic is, in both academia and the practical world, and even the promoters admit that it is rather complex. This limited knowledge so far has subjected ESG to different interpretations, origins, relations, information/data, and even motivations. This introductory chapter on the topic of ESG, provides a definition, traces its history and evolution, and addresses many related factors or issues in an attempt to bring clarity to the topic.

1.2 Definition, Measurement, and Importance

As the name shows, Environmental, Social, and Governance (ESG) refers to three central criteria or categories by which a company's character and sustainability are determined. Through these key groups of factors, the ethical impact of investment in a business can also be determined as well as its future performance (Eccles et al. 2014). Beyond the interests of companies and their shareholders, the goals of ESG are to reverse social,

[2] Brown Brothers Harrison, *2020 Global ETF Investor Survey. ETFs Continue to Chart a New Path*. New York, 2020.

[3] Nina Chestney, "Green Bond Issuance Hit Record $255 Billion Last Year, Research Shows," Reuters, January 16, 2020.

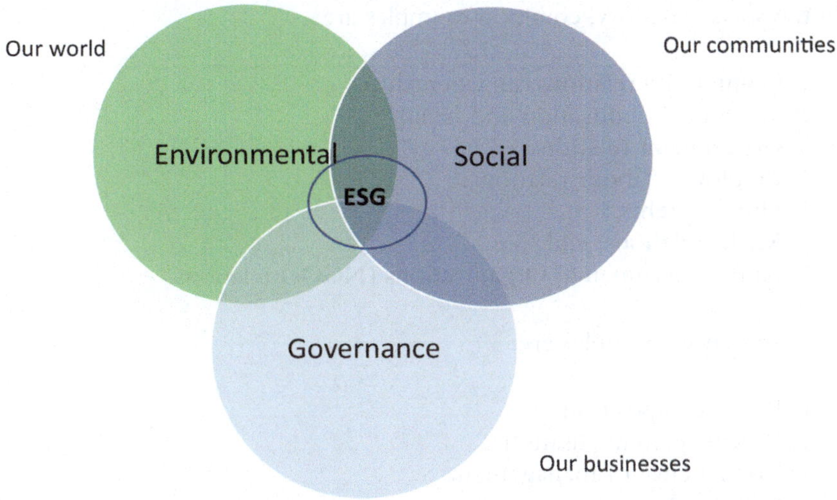

Fig. 1.1 Environmental, social, and governance (ESG) (*Source* Authors)

business, and government actions or inertia that contribute to social and economic inequities and the global climate crisis.[4]

In practice, ESG is implemented by identifying key factors in each of the three categories and their costs or benefits taking into account the activities of the firm or company. For new projects this may be done through environmental impact assessments (EIAs) at the time of conducting the feasibility studies, but for ongoing concerns at later stages.

In the environmental category, the following factors are common examples:

1. Pollution;
2. Waste management;
3. Energy efficiency;
4. Animal welfare; and
5. Plant conservation.

[4] See Chapter 3 for a mathematical analysis with specific regard to the environment and equity within the context of ESG.

In the social category, common examples are:

1. Community relations/engagement;
2. Community education and health;
3. Government relations;
4. Employee/labour relations;
5. Human rights;
6. Media relations; and
7. Non-Governmental Organizations (NGOs) relations.

In governance, examples are:

1. Board composition;
2. Executive compensation;
3. Competence of management;
4. Shareholder relations; and
5. Transparency.

ESG is often confused with the better-known concepts of Corporate Social Responsibility (CSR), Impact Investing (II), and Socially Responsible Investing (SRI). There are many other related terms and names, such as Responsible Investing and Sustainable Investing.[5] This is what causes a lot of confusion as to what ESG really is, where it came from and how old it is. As listing all the related terms is unnecessary, we focus on the "big three" associated terms (CSR, II, and SRI) which are seen by some as competitors and by others as complements to ESG. In any case, they are all subjects of much attention in contemporary financial evaluations and reporting as well as in economic development.

As can be seen from the list of examples of ESG factors above, the big three and ESG are all related, but ESG is broader as it can encompass all the factors and processes of the other three. CSR refers mainly to direct community engagements through community projects and outreach, such as building schools and hospitals, which are intended to give companies their social licences. II on the other hand is focused directly on

[5] https://www.investopedia.com/terms/e/environmental-social-and-governance-esg-criteria.asp.

the beneficiaries of an investment (the recipients) or the positive principle that the investment represents directly and indirectly with regard to the community and/or the environment in which it takes place. For example, an investor may put money into a youth employment fund or an agribusiness company because of the employment and environmental benefits, respectfully, whether or not it makes much profit. SRI is a process through which investments are carried out on the basis of how they conform to an investor's ethical or religious considerations. For example, an investor may refuse to put money into a cigarette or gun manufacturing company, a brewery, casino, or a restaurant that serves pork. Therefore, CSR, II, and SRI are different ways of putting emphasis on various but related ESG factors or issues, all of which are intended to benefit businesses, communities and/or the world at large. The goals are generally the same, although individual motivations, perceptions, and priorities may be different. However, we must reiterate that not everyone sees ESG as encompassing the other three; some explain the relationships being vice versa—that the big three encompasses ESG. Therefore, these relationships are very much similar to that of the chicken and egg. If the egg is dated to the chicken, the chicken becomes older and if the chicken is traced to the egg, the egg becomes older; however, in both cases there is a symbiotic relationship which is questionable mainly by appearance to the eyes or sound to the ears.

Social and regulatory pressures, national and international, are the causes of the shift from elective and optional commitments to mandatory (or what are becoming mandatory) specifications in the conduct of business worldwide (Brockett and Rezaee 2012). Therefore, large listed companies have shifted or are now shifting from publishing roughly one page for employee-related and other disclosures in the 1970s to thorough stand-alone sustainability reports (Gray et al. 1995; Gray Javad et al. 2001). Macroeconomically, corporate sustainability is attained once industrial development subscribes to the three principles of economic efficiency, social equity, and environmental responsibility. At a company level, this can be achieved through a triple bottom line or "three Ps" approach (see Chapters 3 and 7 also). This translates to being able to manage simultaneously a company's risks, obligations, and opportunities within financial, social, and environmental sectors (Elkington 1994). Investors concerned about social responsibility customarily consider ethical, environmental, social, and corporate governance (ESG) aspects when analysing investments, portfolios, and ownership (Galbreath 2013).

Veritably, with the rise of serious environmental and social issues (pollution, waste management, oil leakages, nuclear accidents, water scarcity, child labour, global warming, viruses such as Ebola in 2014 and now COVID-19 [coronavirus], etc.) as well as numerous global governance scandals (such as Enron and Worldcom), investors, and firms are gradually becoming more aware of the necessity of publishing extra and non-financial data (Bassen and Kovacs 2008).[6] Consequently, many firms have increasingly (willingly or unwillingly) decided to employ and disclose ESG programmes.

On the one hand, ESG is voluntarily defined by the environmental, social, and corporate governance performance examined during corporate decision-making processes. On the other hand, ESG information disclosure may be defined by the securities issuer's legal requirements in which enterprises/firms are required or requested to publish ESG and financial management information (Minutolo et al. 2018). This information should be released accurately, on time and comprehensively for the market to make a rational judgement on the value of investments (in line with the Efficient Market Hypothesis). The information disclosure provides a mechanism which assures and safeguards shareholders' and creditors' lawful rights and interests (Fama 2000).

The emphasis on all or any of the three categories (environmental, social, and governance) varies from one company/firm to another. The emphasis on profitability only and/or ESG also differs from one company to another. For some companies, the emphasis is on profitability with no or minimal attention to the three areas of ESG action. For others the emphasis may be on both profitability and ESG. Historically, the former is the case for most, if not all, companies because of the lack of, or minimal, public scrutiny of corporate behaviour and global environmental and other issues. As the public scrutiny and pressures increase, due to factors that would be explained in greater detail later in this chapter, the number of companies for which ESG factors and financial performance become more or less equal would naturally increase. As a result, the need for success in both areas would become more challenging. In fact, the trend has already actively started.

[6] For an extensive list of such scandals and a succinct analysis See Karamo Sonko, "Scandals Bedevil US Corporate Leadership," by Sonko, Dr. Karamo, *African Business*, March 2003.

ESG is rapidly increasing in importance in and outside the world of business. At the current pace, it may soon become legally binding in many countries across the world, especially for big companies. In fact, the volume of regulations concerning ESG has, recently, increased tremendously. According to the London Stock Exchange Group (LSEG), "over 80% of the world's top economies by GDP in 2016 mandated ESG reporting in some form".[7]

> The United Nations organization, with the support of the world's largest institutional investors, has declared Environmental, Social and Governance (ESG) as the first of its Six Principles of Responsible Investing, launched at the New York Stock Exchange in April 2006[8]:
>
> We will incorporate environmental, social and corporate governance (ESG) issues into investment analysis and decision-making processes. Signatories can follow the first principle by supporting the development of ESG-related tools, metrics and analyses and by encouraging research and analysis by service providers and academics on ESG-related issues.

According to the Global Sustainable Investment Alliance (GSIA), the year 2020 showed continuous growth of sustainable investments across regions, with assets under management growing by 15% in two years to

[7] London Stock Exchange Group (LSEG), *Revealing the Full Picture: Your Guide to ESG Reporting. Guidance for Issuers on the Integration of ESG into Investor Reporting and Communication*, 2020, p. 18.

[8] The other principles are:

> Principle 2: We will be active owners and incorporate ESG issues into our ownership policies and practices.
> Principle 3: We will seek appropriate disclosure on ESG issues by the entities in which we invest.
> Principle 4: We will promote acceptance and implementation of the Principles within the investment industry.
> Principle 5: We will work together to enhance our effectiveness in implementing the Principles.
> Principle 6: We will each report on our activities and progress towards implementing the Principles. https://www.unpri.org/pri/what-are-the-principles-for-responsible-investment.

reach US$35.3 trillion or 36% of total professionally managed assets globally (GSIA 2020).[9] Bloomberg Intelligence has forecasted that the market value of ESG investments will increase to $53 trillion in the next five years, representing a third of total assets under management.[10]

With regard to Africa, the continent's only AAA-rated multilateral development finance institution—the African Development Bank (AfDB)—has emphasized that:

> As a leading development finance institution, Environmental, Social and Governance (ESG) issues rank very high on our corporate agenda, particularly in a time of resource constraints and rising social pressures, with public and private organizations facing an increased need for a strong approach to ESG integration in their operations.[11]

Even in Africa, where (as in all less developed countries, generally) ESG disclosures are yet to take root, knowledge about the importance of ESG seems to be increasing, albeit slowly, as shown in news reports. South Africa's *Daily Maverick*, for instance, pointed out that:

> With Africa struggling to meet its sustainable development goals, responsible investment has a key role to play, and ESG (environmental, social and governance) provisions will be crucial in catalysing additional capital flows.[12]

1.3 Origins

The actual ages of ESG and related concepts are the subject of very different explanations, sometimes even speculatory, contradictory, and confusing. For example, Townsend traced socially responsible investing to

[9] The Global Sustainable Investment Alliance (GSIA), *The Global Sustainable Investment Review 2020*. Brussels, Belgium, 2020. See http://www.gsi-alliance.org/wp-content/uploads/2021/08/GSIR-20201.pdf and http://www.gsi-alliance.org/wp-content/uploads/2021/08/GSIR-20201.pdf.

[10] https://www.bloomberg.com/professional/blog/expert-insights-into-data-and-esg-in-the-middle-east/, April 13, 2021.

[11] https://www.afdb.org/en/topics-and-sectors/topics/environmental-social-and-governance-esg/background, August 24, 2021.

[12] *Daily Maverick* on September 28, 2018.

"two millennia", but that they "grew in prominence during the 1980s and 1990s … shaped by civil-rights-era thinkers, faith-based organizations, and women" (2020, p. 10).[13]

Purwar (2019) traced Socially Responsible Investment (SRI) to 200 years ago when the Methodist Church protested against investing in companies that made controversial products such as weapons and tobacco. Then she came back to the 1960s protests, during the Vietnam war, and much more recently to the launching of the United Nations Principles for Responsible Investment (UN PRI) in 2006, after the Freshfield Report[14] and *Who Cares Wins*.[15] She argued that the term ESG was first used in the latter but that it was the former that recommended that ESG issues must be included in financial evaluations. The report made a definitive case for ESG to be considered as a fiduciary duty of managers based on the conclusions of one of the biggest legal firms in the world, Freshfields Bruckhaus Deringer. This moved ESG from the fringes to the mainstream today to become formal comprehensive ESG.

Kelley and Sardi also traced the origins of SRI investing to the 1800s, when the Methodist Church urged members not to invest in alcohol, tobacco, weapons, gambling, and other controversial companies. According to them, ESG became much more mainstream in the 1960s,

[13] Blaine Townsend, "From SRI to ESG: The Origins of Socially Responsible and Sustainable Investing," *The Journal of Impact and ESG Investing*, Vol. 1, No. 1 (Fall 2020): 10–25.

[14] The United Nations Environment Programme (UNEP) FI Asset Management Working Group (AMWG) invited Freshfields Bruckhaus Deringer law firm to assess whether institutional investors, such as pension funds and insurance companies, are legally permitted to integrate ESG into their investment decision-making processes. The Freshfield Report documents their findings after examining the legal framework of these institutions in Australia, Canada, France, Germany, Italy, Japan, Spain, the United Kingdom, and the United States. See *A Legal Framework for the Integration of Environmental, Social and Governance Issues into Institutional Investment*. London: Freshfields Bruckhaus Deringer, 2005.

[15] The latter was an initiative of the UN Secretary General and UN Global Compact in cooperation with the Swiss government in 2004, endorsed by 23 financial institutions with total assets of over US$6 trillion. See IFC Advisory Services in Environmental and Social Sustainability, *Who Cares Wins, 2004–08*, Issue Brief 2004. https://www.ifc.org/wps/wcm/connect/9eeb7982-3705-407a-a631-586b31dab000/IFC_Breif_whocares_online.pdf?MOD=AJPERES&CACHEID=ROOTWORKSPACE-9eeb7982-3705-407a-a631-586b31dab000-jkD12B5#:~:text=Who%20Cares%20Wins%20(WCW)%20was,US%246%20trillion%20in%20assets.

aided by the growth of the mutual fund industry, the civil rights movement, and the protests against companies involved in the Vietnam War. They argued that in the 1980s, ESG investing helped to end apartheid in South Africa when churches, universities, cities, and states during the period made decisions which forced American companies to divest in South Africa, causing very serious economic instability in the country.

However, they admitted (as we argue in this book) that although the idea of restricting investments based on values or ethics is nothing new, it was not until recently that ESG investing emerged to include a large range of strategies, issues, and styles that cover all asset classes and regions. They believed that "Present-day ESG investing essentially took hold first in Europe, where regulations and standards relating to ESG started to develop in the early 2000s".[16]

According to Atkins (2020) too, ESG issues were first stated in 2006 by the UN in its PRI report. ESG reporting was seen as the means of increasing sustainable investments, she argued.[17] At the time, 63 asset owners and service providers with $6.5 trillion in assets under management signed an agreement to adopt ESG disclosures. As of June 2019, the number of companies which became signatories to this agreement reached 2450 with more than $80 trillion of assets under management.[18]

The World Bank traced the history of ESG as follows:

> ESG investing has been gathering attention since the 1990s. From its origins in the equity markets with religious, values-based or thematic (environmental) investors, the movement spread with the launch of the UN Principles of Responsible Investment (UNPRI) in 2006 and was catalyzed for fixed income with the issuance of labelled bonds by multilateral organizations from 2007. (The World Bank 2018)[19]

The religious momentum behind ESG investing in recent years was gathered mainly from the emergence of formal Islamic finance, which has

[16] https://www.esl.org/resources-tools/educational-resources/esg-investing-part-one, August 24, 2021.

[17] Cited above in footnote 14.

[18] Betsy Atkins, "Demystifying ESG: Its History & Current Status," *Forbes*, June 8, 2020.

[19] Georg Inderst and Fiona Stewart, *Incorporating Environmental, Social and Governance (ESG) Factors into Fixed Income Investment*. World Bank Group, 2018, p. 1.

grown very rapidly since the 1990s. Over the decade of the 2000s, the Islamic finance industry grew by more than 10% annually to reach an estimated $1.3 trillion by 2011. Total assets were about $1.9 trillion in 2016, reached $2.4 trillion in 2017, and were forecasted to reach $3.2 trillion by 2020.[20]

Several factors contributed to this growth of Islamic finance: (1) the increasing need of Muslims for *Shari'ah*-compliant products and non-Muslims for ethical financial products; (2) petrodollar surpluses in the Middle East; (3) development of regulatory frameworks in Muslim and non-Muslim countries; (4) increasing demand for large-scale infrastructure financing; and (5) global financial crises which have raised questions about the resilience of the conventional financial system.[21]

Although the World Bank and some of the others cited above have traced its origins to the 1990s or earlier, ESG as a formal comprehensive or holistic concept certainly emerged much later. From our literature survey, the earliest evidence of this we can find is 2006, reported by Atkins and Purwar (above). As illustrated before, the last decade and half (that is, since the UN PRI (Eccles and Klimenko 2019), in particular, witnessed an almost dramatic surge in interest in, and concerns about, the state of the global environment, sustainability, corporate governance, and the community/local benefits of large investment projects. These led to the evolution of what we refer to as formal comprehensive or holistic ESG, due to a combination of factors, all of which we attempt to outline in this chapter.

[20] Karamo N. M. Sonko, and Mariama Sonko, "Promoting Islamic Finance in Africa: Top–Bottom or Bottom-Up? A Case of Islamic Banking in the Gambia," in *Islamic Finance as a Closed System: New Insights*, ed. Haider A. Khan, and Karamo N. M. Sonko. Lanham, MD: Lexington Books, 2020.

[21] For further information on Islamic finance see Haider A. Khan, and Karamo N. M. Sonko, *Islamic Finance as a Closed System: New Insights*. Lanham, MD: Lexington Books, 2020.

1.4 Actors

1.4.1 The United Nations

As a result of formal public scrutiny, calls for voluntary, if not obligatory, ESG reporting have in recent years found their greatest momentum within the United Nations (UN) system and Non-Governmental Organizations (NGOs). The UN perhaps deserves the main credit for setting the scene and kickstarting the efforts, through the establishment of the United Nations Environmental Programme (UNEP) in 1972. The UNEP is a symbol of not only the UN's environmental concerns but also global social concerns which the UN has galvanized over the years, resulting in an increasingly strong relationship with NGOs, and now businesses.

The establishment of the UNEP and the United Nations Conference on the Human Environment, in Stockholm the same year, were expressions of growing worldwide environmental concerns.[22] UNEP's attempts to find solutions for environmental problems, such as pollution in the Mediterranean Sea, the threat to aquatic resources, deforestation, desertification, and drought, the depletion of the Earth's ozone layer through chemicals and the consequent global warming, have attracted much attention. Through the General Assembly, UNEP succeeded in setting up the World Commission on Environment and Development.

The 1992 Rio "Earth Summit" is particularly significant in boosting the UN's and global efforts, because it resulted in many environmental initiatives and agreements, including the United Nations Framework Convention on Climate Change (UNFCCC), which is governed by the UN's Conference of Parties (COP). The Earth Summit, the biggest intergovernmental global conference of its kind ever, resulted in the Convention on Biological Diversity, the Global Warming Convention, the Rio Declaration on Environment and Development, the Statement of Principles on Forests, and the Plan for the Sustainable Development of the Earth's Resources into the twenty-first century (known as Agenda 21). Subsequent large-scale conferences and landmark global accords under the auspices of the UN are the: Kyoto Protocol (1997); Bali Action Plan (2007); Copenhagen Accord (2009); Cancun Agreements (2010); Durban Platform for Enhanced Action (2011); and Paris Agreement

[22] This was the first UN conference to make the environment a global concern. It resulted in a set of adopted environmental principles and resolutions, including the Stockholm Declaration and Action Plan for the Human Environment.

(2015). The UN's ability to bring together the world's political leaders has boosted and given credibility to its efforts and those of civil society organizations associated with it.

The Earth Summit incorporated inputs from numerous Non-Governmental Organizations (NGOs) (which became its strongest supporters), and other major contributors to global environmental campaigns and corporate social responsibility. Many NGOs are now members of UN bodies or associations formed by the UN.[23] Whilst developed country governments are the frontrunners in calling for and ensuring proper corporate governance, NGOs are arguably the frontrunners in pushing for CSR and social awareness or community consciousness in areas where businesses, especially multinational companies in natural resources or the extractive industries, operate. The emergence and popularity of the concepts of the "paradox of plenty or the resource curse" and NGOs, such as Global Witness and Publish What You Pay, have highlighted the issues which put the extractive industries at the forefront of pressures for attention to environmental, social, and governance issues by the corporate sector even before what became formal ESG. These issues led to the launching of the Extractive Industries Transparency Initiative (EITI) in 2002 at the World Summit on Sustainable Development. Therefore, such developments were part of the narrative that led to the emergence of ESG and are extensively discussed in this chapter.

The Millennium Development Goals (MDGs) were eight international development goals set by the UN for the year 2015 after the Millennium Summit in 2000 and the adoption of the Millennium Declaration. These were:

1. Eradication of extreme poverty and hunger
2. Achievement of universal primary education
3. Promotion of gender equality and empowerment of women
4. Reduction of child mortality
5. Improvement of maternal health
6. Combatting HIV/AIDS, malaria, and other diseases
7. Ensuring environmental sustainability
8. Development of a global partnership for development.

[23] For example, TI has UNESCO Consultative Status, is a member of the United Nations Global Compact and is closely associated with the United Nations Sustainable Development Group (UNSDG).

The Sustainable Development Goals (SDGs) succeeded the MDGs in 2016 and comprises 17 Sustainable Development Goals (SDGs) and 169 targets set by the United Nations General Assembly in 2015 (Fig. 1.2). These goals were targeted for achievement by the year 2030. They are:

1. No poverty
2. Zero hunger
3. Good health and well-being for people
4. Quality education
5. Gender equality
6. Clean water and sanitation
7. Affordable and clean energy
8. Decent work and economic growth
9. Industry, Innovation, and Infrastructure
10. Reducing inequalities
11. Sustainable cities and communities
12. Responsible consumption and production
13. Climate action
14. Life below water

Fig. 1.2 The UN sustainable development goals (*Source* https://www.un.org/sustainabledevelopment/. This figure has been officially approved by the UN but the contents of this publication have not been approved by the UN and does not reflect the views of the UN or its officials or Member States)

15. Life on land
16. Peace, justice, and strong institutions
17. Partnerships for the goals.

The UN Global Compact was officially launched in 2000 (same year as the MDGs). It is a non-binding agreement to encourage businesses to adopt and implement sustainable and socially responsible practices, and to report on them. The Global Compact is the world's largest corporate sustainability initiative with 13,000 corporate participants and other stakeholders from more than 170 countries. It has two objectives: "Mainstream the ten principles in business activities around the world"[24] and "Catalyse actions in support of broader UN goals, such as the Millennium Development Goals (MDGs) and Sustainable Development Goals (SDGs)".[25]

The efforts of the Global Compact from 2004 to 2009 led to the establishment of the Sustainable Stock Exchange Initiative (SSEI) in the latter year (Fig. 1.3). This is a Partnership Programme launched by the Global Compact, United Nations Conference on Trade and Development (UNCTAD), United Nations Environmental Program Financial Initiative (UNEP FI), and PRI. Its mission is to provide a platform for stock exchanges, working with investors, companies, regulators, policymakers, and international organizations, to pursue ESG objectives, enhance sustainable investment, and the financing of the UN Sustainable Development Goals.

1.4.2 Governments

In addition to the efforts of the United Nations organization and NGOs, governments in the developed countries, especially Canada, the United Kingdom, and the United States, have *directly* helped to regulate the behaviour of their corporations at home and abroad. In Canada, the Extractive Sector Transparency Measures Act (ESTMA) was enacted in

[24] The Ten Principles are derived from: the Universal Declaration of Human Rights, the International Labour Organization's Declaration on Fundamental Principles and Rights at Work, the Rio Declaration on Environment and Development, and the United Nations Convention Against Corruption. For the list visit https://www.unglobalcompact.org/what-is-gc/mission/principles.

[25] https://en.wikipedia.org/wiki/united_nations_global_compact.

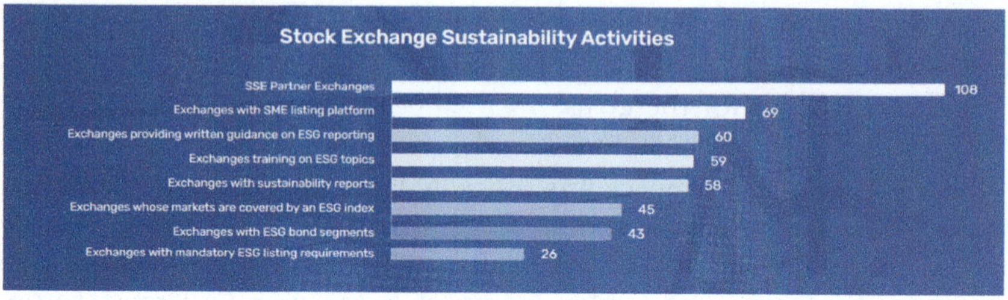

Fig. 1.3 Stock exchange sustainability activities (*Source* https://sseinitiative.org/)

2014, and brought into force on 1 June 2015. It aims "to contribute to global efforts to increase transparency and deter corruption in the extractive sector by requiring extractive entities active in Canada to publicly disclose, on an annual basis, specific payments made to all governments in Canada and abroad".[26] With regard to the United Kingdom, the Extractive Industries Transparency Initiative (EITI) in 2002 (which we have referred to above), for example, was actually launched at the World Summit on Sustainable Development in Johannesburg by the then British Prime Minister Tony Blair. In the United States, the Foreign Corrupt Practices Act (FCPA) of 1977 (comparable to the Bribery Act in the United Kingdom) holds American companies legally responsible for the behaviour of their distributors and suppliers globally. However, it is only in recent years that it came to be enforced.

In short, the United Nations, NGOs, and governments, outside the multilateral system, have all helped to highlight issues which led to the rise of CSR, II, SRI, and finally formal comprehensive or holistic ESG. The environment has attracted the greatest attention because of global warming. Although not everyone subscribes to it, the concern about it is almost universal. Even development financial institutions, such as the African Development Bank and the International Financial Corporation (IFC) of the World Bank Group, as well as private banks which finance large-scale projects, now insist on environmental impact assessments (EIA) as conditions for their financing.

[26] https://www.nrcan.gc.ca/mining-materials/resources/extractive-sector-transparency-measures-act-estma/18180.

The UNEP FI has directly and publicly supported and recommended the integration of some of what are today known as ESG-related factors into the processes of corporate decision-making since 1992. Pursuant to such initiatives, in May 2004, the Johannesburg Stock Exchange (JSE) became the first African exchange to launch the Socially Responsible Investments (SRI) Index in order to distinguish and facilitate investment in listed companies that incorporate the triple bottom-line activities approach to business (Labuschagne et al. 2005). Such activities of the JSE are elaborated in Chapter 7.

1.4.3 Non-Governmental Organizations (NGOs)

The rise of NGOs has been explained through theoretical perspectives which are influenced by their origins in the developed economies. The dominant approach explains the phenomenon in terms of the failure of the market and the state. This approach focuses on subsidies, public goods, contracts, and consumer–producer relationships which justify the need for NGOs. In summary, these theories argue that NGOs emerge because of subsidies to them and the failure of the state and market to satisfy the needs of civil society or the population. Adapted to ESG, this means that NGOs focus on environmental, social, and governance issues because of the failure of governments and businesses to address these concerns. Therefore, NGOs exist to fill the gap created by the inability or unwillingness of the two sectors (public [government/state] and private [business]) to meet their national and international obligations.

NGOs fall into different categories, which cut across issues covered by the environment, society, and governance. There are those that are considered radical or somewhat militant, such as Green Peace, which tend to focus on the environment. There are moderate ones such as Amnesty International and Human Rights Watch, which focus on human rights, and others, such as Oxfam and Save the Children, which focus on economic issues (poverty) and matters affecting children or other vulnerable groups, respectively. Another category of NGOs, such as Transparency International (TI), focuses on corruption. The common characteristic of NGOs is that, generally, they do not have the political inhibitions of governments and political organizations (including the UN itself and political parties) and the bias of interest groups, such as business, labour and professional organizations or associations. However, they sometimes come under criticisms for either being too idealistic or unrealistic and for raising funds that they spend more on themselves than the causes they claim to fight for.[27]

Common international goals have strengthened the association between NGOs and UN agencies, in spite of the political limitations of the later imposed by member states, particularly the most powerful states. These common goals are clearly illustrated in the Millennium Development Goals (MDGs) and the Sustainable Development Goals (SDGs) of the UN. Both sets of goals are strongly supported by NGOs.

1.4.4 The Voices of Businesses

In recent years, business persons and corporations, especially from Europe and the United States, have decided to add their voices, in the calls for sustainability. For example, State Street Global Advisors (SSGA) in 2017 voted against the re-election of board members of 400 companies which they said were not progressing in appointing female directors. Exxon Mobile's shareholders in the same year also showed their displeasure when 62% of them voted to force the oil and gas giant to disclose the impact of its activities on the environment.

In August 2019, after more than two decades, the U.S. Business Roundtable overturned its policy statement emphasizing that the main purpose of a company is to maximize shareholder financial value (return). In its place, the 181 chief executive officers (CEOs) of the Roundtable declared a new statement that the purpose is to also serve its customers, employees, suppliers and the community. The Roundtable is an association of the CEOs of the leading American companies. For 22 years, its formal statement on the purpose of the corporation expressed it as

[27] For studies on the UN-NGO relationships, which are sometimes stormy, depending on the kind of NGO, the interested reader may refer to: P. Willets (ed.), *The Consciences of the World: The Influence of Non-Governmental Organizations in the UN System*. London: Hurst, 1996; T. Princen, and M. Finger (eds.), *Environmental NGOs in World Politics: Linking the Global and the Local*. London: Routledge, 1994; M. Rech, and K. Sikkink, *Activists Without Borders: Advocacy Networks in International Politics*. Ithaca, NY: Cornell University Press, 1998; Bas Arts, Math Noortmann, and Rob Reinalda (eds.), *Non-State Actors in International Relations*. Aldershot: Ashgate, 2001; and William De Mars, *NGOs and Transnational Networks*. London: Pluto Press, 2005. For a collection of papers on the theories, roles, potential, and limits of NGOs in development, see for example, Paul Samuel, and Arturo Israel (ed.), *Nongovernmental Organizations and the World Bank Cooperation for Development*. Washington, DC: The World Bank, 1991.

the need to serve shareholders. Therefore, according to this statement, it should now be expected to focus not on the need to serve shareholders, but pay attention to the interests of other stakeholders too.

The International Business Council (IBC), in partnership with the Big Four accounting firms (Deloitte, PwC, KPMG, and Ernst & Young) and the World Economic Forum (WEF) are pushing forward the ESG agenda and working on developing standardized metrics for disclosures/reporting. The IBC comprises 120 of the largest multinational corporations in the world. Their efforts contributed to the *Davos Manifesto of 2020: The Universal Purpose of a Company in the Fourth Industrial Revolution*, declared at the 50th Anniversary Meeting of the WEF in Davos, Switzerland, 21–24 January 2020. The calls and activities for the Manifesto, were headed by Klaus Schwab, Founder and Executive Chairman of the World Economic Forum. The intellectual arguments for the Manifesto are contained in a book published by Schwab and his head of communications, Peter Vanham, in 2021.[28]

The Davos Manifesto of 2020 listed, systemically, 22 core metrics and 34 expanded metrics that are market-driven and based on existing frameworks (see Figs. 1.4 and 1.5a–d).[29] They are also aligned with the UN's Sustainable Development Goals (SDGs). The metrics were grouped and developed in four critical areas by the Big Four: (1) the principles of governance (Deloitte); (2) planet (PwC); (3) people (KPMG); and (4) prosperity (Ernst & Young).

The 2020 Manifesto is actually the second such manifesto of WEF. The first one was launched as far back as in 1973. The 2020 Manifesto states that: "The purpose of a company is to engage all its stakeholders in shared and sustained value creation. In creating such value, a company serves not only its shareholders, but all its stakeholders – employees, customers, suppliers, local communities and society at large. The best way to understand and harmonize the divergent interests of all stakeholders is through a shared commitment to policies and decisions that strengthen the long-term prosperity of a company".

[28] Klaus Schwab, and Peter Vanham, *Stakeholder Capitalism: A Global Economy that Works for Progress, People and Planet*. Hoboken, NJ, 2021.

[29] For a complete report of these see World Economic Forum in collaboration with Deloitte, EY, KPMG and PwC, *Toward Common Metrics and Consistent Reporting of Sustainable Value Creation*. Davos, Switzerland, January 2020.

Pillar	Theme	Sub theme	Source
Principles of Governance	Governing Purpose	Setting purpose	GRI (102-26), EPIC, Colin Mayer and others
	Quality of Governing Body	Board composition	GRI (102-22), GRI (405-1a)
	Stakeholder Engagement	Impact of material issues on stakeholders	GRI 102-47
	Ethical behaviour	Anti-corruption	Adapted from GRI (205-2) and GRI (205-3)
		Protected ethics advice and reporting mechanisms	GRI (102-17)
	Risk and Opportunity Oversight	Integrating risk and opportunity into business process	Combination of EPIC and SASB (230a.1 and 2)
Planet	Climate Change	Greenhouse Gas (GHG) emissions	GRI (305-1), CDP (C6, C7), CDSB (R03, R04), SASB (110a.1), GHG Protocol
		TCFD-aligned reporting on material climate risks and opportunities	TCFD, CDSB R01, R02, R03, R05 and R06; SASB 110
	Nature Loss	Land use and ecological sensitivity	Adapted from: GRI (304-1, 304-3, 304-4), CDP (F1)
	Fresh Water Availability	Fresh water consumption in water stressed areas	Adapted from: GRI (303-3), CDP (W1), CDSB (R04), SASB (140a.1)
People	Dignity and Equality	Gender pay equality (%)	GRI (405-2)
		Diversity and inclusion (%)	GRI (406-1)
		Wage level (%)	GRI (202-1)
		Risk for incidents of child, forced or compulsory labor (#, %)	GRI (408, 409)
	Health and Well Being	Health and safety (%)	SASB (CN010118), GRI (403-2.a4)
	Skills for the Future	Training provided (#)	GRI (404-1), SASB (HC0101-15)
Prosperity	Wealth creation and employment	Net number of jobs created	GRI (401-1a & b)
		Net Economic Contribution	GRI (201-1 and 201-4)
		Net investment	International Accounting Standard (IAS) 7 – Cash Flow Statements
	Innovation in better products and services	R&D spend ratio (%)	2015 edition of the Frascati Manual for measuring R&D (OECD, 2015a)
	Community and social vitality	Community investment (%)	GRI (G4-ECI)
		Country by country tax reporting	GRI (207-4)

Fig. 1.4 Summary overview of core metrics and disclosures (*Source* Authors, extracted from WEF, Deloitte, EY, KPMG, and PWC, ESG Metrics Discussion Paper, pp. 7–8)

a

Pillar	Theme	Expanded Metrics and Disclosures	Sources
Principles of Governance	Governing Purpose	Material stakeholder buy-in	New metric
	Quality of Governing Body	Progress against strategic milestones	EPIC
		Remuneration	GRI (102-35)
	Stakeholder Engagement	Process for engaging stakeholders	GRI (102-43)
	Ethical Behaviour	Alignment of strategy and policies to lobbying	CDSB (REQ-01)
		Monetary losses from unethical behaviour	SAM (3.4.5); SASB (510a.1)
	Risk and Opportunity Oversight	ESG in capital allocation framework	CDSB (REQ-01)

b

Pillar	Theme	Expanded Metrics and Disclosures	Sources
Planet	Climate Change	Science-based target to reduce GHG emissions	SBTi
		TCFD-aligned reporting	TCFD
		Impact of greenhouse gases	NCP, ISO 14008
	Nature Loss	Impact of land use	NCP, ISO 14008
	Fresh water availability	Impact of fresh water consumption	NCP, ISO 14008
	Air pollution	Fine particulate matter	GRI (305-7), SASB (120a.1)
		Impact of air pollution	NCP, ISO 14008
	Water pollution	Nutrients	Adapted from: GRI (303-1)
		Impact of water pollution	NCP, ISO 14008
	Solid waste	Single use plastics	New Metric
		Impact of solid waste disposal	NCP, ISO 14008
	Resource availability	Resource circularity	WBCSD & KPMG Circular Transition Indicators

Fig. 1.5 **a** Summary overview of expanded metrics and disclosures, Pillar 1—Principles of Governance (*Source* Extracted from WEF, Deloitte, EY, KPMG, and PWC, ESG Metrics Discussion Paper, p. 15); **b** Summary overview of expanded metrics and disclosures, Pillar 2—Planet (*Source* Authors, extracted from WEF, Deloitte, EY, KPMG, and PWC, ESG Metrics Discussion Paper, pp. 17–18); **c** Summary overview of expanded metrics and disclosures, Pillar 3—People (*Source* Authors, extracted from WEF, Deloitte, EY, KPMG, and PWC, ESG Metrics Discussion Paper, p. 21); **d** Summary overview of expanded metrics and disclosures, Pillar 4—Prosperity (*Source* Authors, extracted from WEF, Deloitte, EY, KPMG, and PWC, ESG Metrics Discussion Paper, p. 24)

c

Pillar	Theme	Expanded Metrics and Disclosures	Sources
People	Dignity and Equality	Discrimination and harassment incidents (#) and the total amount of monetary losses ($)	GRI 406-1, SASB FB-FR-310
		Freedom of association and collective bargaining (%)	SASB CN0401-17, GRI 407 WDI 7.2
		Living wage (%)	MIT Living Wage Tool, EPIC Report, Shift
		Grievances and impact (#, $)	UN Guiding Principles Reporting Index, GRI WDI 7.5
	Health and Well-being	Monetized impacts of work-related incidents on employees, employers and society (#, $)	Adapted indicator, based on European Commission, Safe Work Australia
		Well-being (%)	Embankment Project
	Skills for the Future	Number of unfilled 'skilled' positions (#)	WBCSD Impact Framework
		Monetized impacts of training – increased earning capacity as a result of training intervention ($)	OECD, United Nations

d

Pillar	Theme	Expanded Metrics and Disclosures	Sources
Prosperity	Employment and Wealth Generation	Average wage	SASB (FB-RN-310a.2)
		Significant indirect economic impacts	GRI (203-2)
	Innovation of Better Products and Services	Vitality Index	OECD Oslo Manual Section 8.3.1
		Net Promoter Score (NPS)	Reichheld, Bain & Company and Satmetrix, Harvard Business Review
		Social value generated (%)	Adapted from SASB FN0102-16.a and GRI (FiFS7 + FiFS8)
	Community and Social Vitality	Infrastructure investments and services supported	GRI (203-1)
		Total social investment ($)	CECP 2020 Valuation Guide

Fig. 1.5 (continued)

Therefore, the World Economic Forum has moved to the forefront of the international advocacy for ESG by bringing together influential businesses and business organizations discussed above. In addition, its Chairman has coined a "new" concept referred to as "stakeholder capitalism", which is gaining ground in reinforcing the Forum as an arena for ESG advocacy. In the book cited above, Schwab and Vanham observed that the world is facing very serious structural problems, viz.:

1. The challenge posed by rising income inequality, even in economies where productivity and wage growth have slowed down.
2. Increasing market concentration in various industries, and the related issues of slowing growth, innovation, and productivity gains, as well as higher indebtedness.
3. The short-sighted exploitation of natural resources, that is corroding the environment and affecting the lives of many people for the worse (also refer to Khan and Sonko, 1994 in Chapter 3 for a related point demonstrated mathematically).

These, they argued, are caused by globalization, the Fourth Industrial Revolution, and the misuse of natural resources. Ironically, they pointed out that, the causes of our prosperity are also often the causes of the problems around the world.

The authors asked and answered the following questions:

1. How can we build a global economy that works for progress, people, and the planet?
2. Where did the economic system that prevailed in the post-war era succeed, and where does it need a course-correction?
3. And how can we avoid the pitfalls posed by advances in technology and global trade, whilst harnessing their strengths?
4. How can a more inclusive, sustainable, and resilient global economy, post-COVID, be achieved.

They provided the answers in what they described as a "new" form of capitalism which they referred to as "stakeholder capitalism". Unlike shareholder and state capitalisms, their model calls for the integration of the interests of all stakeholders in a company's business. These stakeholders, in addition to the company itself (shareholders and employees), are the government, community, and larger society. The authors argued that collaboration between them, long-term visions as opposed to short-term goals, and measures of success broader than financial profits and GDP are the cornerstones of this form of capitalism, which can help the world achieve resilience, inclusiveness, and sustainability in a global economy that serves progress, people, and the planet.[30]

1.4.5 *The Stock Exchanges*

The impact of the voices of businesses, discussed above and shown in what Ernst & Young Canada (2021) described as the "record flow of ESG funds",[31] Wyatt (2021) called the "rise and rise of ESG",[32] and Kell (2018)[33] called the "remarkable rise of ESG" can be seen in the stock markets of the world, even in Africa (see Chapter 6). On their own, and in collaboration with the UN agencies shown earlier, the major stock exchanges of the world, including the top ten markets New York, National Association of Securities Dealers Automated Quotations (NASDAQ), Tokyo Stock Exchange (TSE), Shanghai Stock Exchange (SSE), Euronext, Hong Kong Stock Exchange (HKSE), London Stock Exchange (LSE), Shenzhen Stock Exchange (SSE), Toronto Stock Exchange (TSE), and BSE (Bombay Stock Exchange), all require some aspect of ESG reporting now.

What this has done is to indirectly force issuers and investors to pay (more) attention to ESG reporting in buying and selling securities—stocks, bonds, and other financial instruments of public companies. In 2019, NASDAQ, the biggest stock exchange in the world, reported that:

[30] Stakeholder Capitalism was published on 27 January 2021. The authors carried out a number of country case studies and concluded that stakeholder capitalism is akin to the Living Standard Framework of the New Zealand government. This seeks to achieve the different aspects of the well-being of citizens in the short and long term, including distributional issues and resilience and are in harmony with the metrics of stakeholder capitalism. For more information on this Framework see https://www.treasury.govt.nz/information-and-services/nz-economy/higher-living-standards/our-living-standards-framework.

[31] Jean-François Gagnon, with Martin Leroux, Alain Deschênes and Aida Chakri as contributors, "The Rise of ESG Investing," *Canadian Asset Manager ESG Survey Report*. Ernst & Young Canada, March 31, 2021.

[32] Helen Wyatt, "Sustainable Finance: The Rise and Rise of ESG," *Lexology*, June 11, 2021.

[33] Robert Kell, "The Remarkable Rise of ESG," *Forbes*, July 11, 2018.

"Nasdaq updated the name of its Nominating & Governance Committee to the Nominating & ESG Committee to better align with its expanded responsibilities".[34] In May 2020, the U.S. Securities and Exchange Commission (SEC) developed an ESG reporting framework through its investment committee. In the same year, the London Stock Exchange Group pointed out in its guide on ESG:

> Once upon a time, environmental, social and governance (ESG) factors were a niche interest among asset owners, asset managers, banks, brokers and investment consultants. No longer. Investors now routinely analyse information on ESG performance alongside other financial and strategic information in order to gain a better understanding of companies' future prospects.[35]

Effectively the stock exchanges are responding to their stakeholders, particularly in the developed countries, and moving in a direction which may ultimately carry an explicitly strong message to issuers: "ESG or no money!"[36] For the sustainability activities of stock exchanges, refer to Fig. 1.3.

1.5 Other Factors in the Recent Evolution of ESG

1.5.1 Natural and Man-Made Disasters

In a world that seems to be increasingly prone to natural and man-made disasters, private companies have come under increasing scrutiny and pressure to participate in the mitigation of the consequences of national and international emergencies. From the Tsunami in Asia to the "9/11" terror attack in the United States, companies cannot stand by anymore, leaving governments and civil society to deal with such large-scale havocs. The response of companies to the COVID-19 pandemic provides a good example of corporate social responsibility and actions which indirectly or

[34] *Nasdaq 2019 Sustainability Report*, p. 6.

[35] London Stock Exchange Group, Revealing the full picture: Your guide to ESG reporting, London, 2020.

[36] Robert G. Eccles, and Svetlana Klimenko, "The Investor Revolution," *Harvard Business Review* (May–June 2019).

directly popularizes the wider and formal concept of ESG. In this context, South Africa, for example, provides a practical and clear evidence of how disasters can lead to, and highlight the importance of, ESG. As shown in Figs. 1.6 and 1.7, the country has the highest number of confirmed COVID-19 cases and deaths on the African continent. On 23 March 2020, President Ramaphosa made a challenge to his country to unite and set up a COVID-19 national Solidarity Fund of R4 billion (US$214 million) in order to:

1. Detect and Prevent
   To understand the magnitude of the infection of COVID-19 and support measures to flatten the curve by lowering infection rates;
2. Care
   To assist with the management of those people in hospital or requiring medical care;
3. Support
   To support those people whose lives are disrupted by COVID-19.

The government contributed seed capital of R150 million to launch the Fund, expecting most of the rest to come from businesses. In fact, of the total of R1.8 billion received by 4 May 2020, R1.3 billion

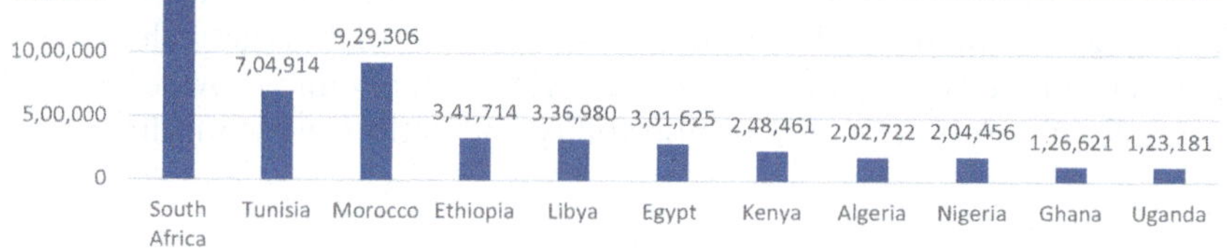

Fig. 1.6 Africa's leading Coronavirus-affected Countries (numbers of confirmed cases) (*Source* Authors, based on Johns Hopkins Coronavirus Resource Center, data for 27 September 2021, 4.14 a.m., GMT. Website: https://coronavirus.jhu.edu/map.html)

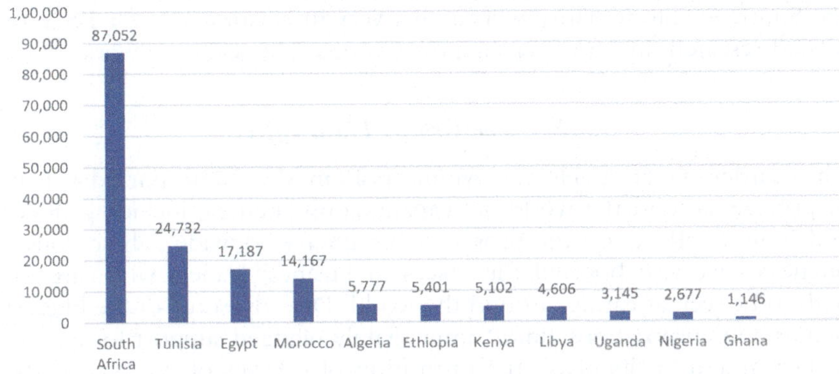

Fig. 1.7 Africa's leading Coronavirus-affected Countries (numbers of confirmed deaths) (*Source* Authors, based on Johns Hopkins Coronavirus Resource Center, data for 27 September 2021, 4.14 a.m., GMT. Website: https://coronavirus.jhu.edu/map.html)

or 72% came from Mary Oppenheimer and Daughters (R1 billion), Naspers Limited (R200,000,000), Lottery South Africa (R50,000,000), and Telesure Investment Holdings (R50,000,000).[37]

Although public–private partnerships for profit-making projects are now common throughout Africa, this Fund is the first not-for-profit public–private partnership of its kind. It directly demonstrates the importance of CSR and indirectly ESG (although without that name) because of the involvement of private corporations. It also highlights the importance of public–private partnership in ESG (which we call for in this book). The National Solidarity Fund is the biggest Fund of its kind ever, by an African government in partnership with private businesses to address the "social" in ESG, with many implications for "environmental" and corporate "governance". Together with the contributions private companies have also put into international foundations, the World Bank and the United Nations agencies for COVID-19 responses around the world,

[37] According to the IMF, 1 United States Dollar equals 18.70 South African Rand, https://www.imf.org/external/np/fin/data/rms_rep.aspx.

the South African initiative will carve a very interesting area for ESG, or related research, in Africa, in the post-Coronavirus academic world.[38]

1.5.2 The George Floyd Effect

The murder of George Floyd in Minneapolis in May 2020 spurred a chain of protests around the world with many consequences, including direct and indirect effects on the prospects for ESG worldwide. These consequences have also boosted the voices of business leaders who care, or wish to be seen as caring, around the world. Directly, some of the biggest brands and corporations almost immediately pledged support for an end to racism and, in its place, the institution of policies of diversity at the board, management, and worker levels. Some companies were very vocal and public in their pronouncements and condemnations.

Indirectly, the murder of Floyd helped to accelerate programmes of diversity which were already in place, being contemplated or ignored in many companies. In May 2021, McDonald's announced that it would increase its national spending on advertising from two to five per cent during the 2021–2024 period, and to ten per cent the spending on minority-owned businesses, including those owned by Asian Americans, Pacific Islanders, and women. In June 2021, the membership of the U.S. Business Roundtable came together and pledged to eliminate all barriers to equal opportunities for all Americans in the economy, and to advance

[38] Such partnership helps avoid the type of potential conflict that is brewing between some private corporations and governments, because of arbitration claims by the latter using investment treaties which "often result in large payouts, sometimes of billions of dollars, to these companies for alleged lost profits. These suits pose an immediate danger to the ability of developing nations, and the global community as a whole, to confront the COVID-19 challenge". Therefore, under the auspices of the Columbia Center on Sustainable Investment and partner organizations, Phil Bloomer, Juan Pablo Bohoslavsky, Carlos Correa, Olivier De Schutter, Kerry Kennedy, Jeffrey D. Sachs, and María Fernanda Espinosa Garcés called "for a complete moratorium on all ISDS (investor-state cases) claims until the pandemic has passed and governments have agreed on principles to ensure that future arbitration cases will not hinder countries' good faith recovery efforts. These principles should recognize the full and clear scope that governments have, and are required, to take all appropriate actions to save lives and fight global emergencies, even when the result is a loss of profits or business opportunities, including by foreign investors". For more information see http://ccsi.columbia.edu/2020/05/05/isds-moratorium-during-covid-19/?utm_source=CCSI+Mailing+List&utm_campaign=3dca9f0160-GRASFI+2020+Moving+Online_COPY_02&utm_medium=email&utm_term=0_a61bf1d34a-3dca9f0160-62904933.

racial equity and justice for minorities. Amazon launched a Black Business Accelerator with a $150-million fund for supporting equity and sustainable growth for black-owned businesses. In June too, Best Buy promised to spend $1.2 billion on Black, Indigenous, People of Color (BIPOC) and other minority-owned businesses by 2025, in order to create a more diverse supplier base and to help increase BIPOC presence in the technology field. Similarly, Coca-Cola North America in June 2021 expressed commitment to increase its spending almost two-fold on minority-owned media companies in the three years that would follow. In the same month, General Mills, Target, Tyson Foods, Uber, Adidas, and 15 other corporations pledged support for GroupM's Media Inclusion Initiative to expend at least two per cent of their total media budgets, per annum, on black-owned ones.

1.5.3 The Generational Effect

Finally, among the contributors to the rise of ESG, is an effect we describe as the "generational effect", which is currently taking place and is happening in favour of ESG. This effect is observed from the two generations comprising Y (Millenials) and Z; in order words, since the so-called Generation X. These young men and women born to the "baby boomers" and the early Generation X have some significantly different orientations from their parents. These orientations tend to include environmental activism, social engagement through electronic media, very self-assertive and vocal. Research has shown that these generations do not tend to like working, investing, or shopping in companies they consider to be out of line with what they believe in—business outlooks broader than maximizing shareholder value, especially with regard to the environment. The clearest expression of this is what has been referred to as the "Greta effect". The young Swedish environmental activist who has described the present global economic system as one that is perpetuating the betrayal of future generations through unsustainable and entrenched environmental malpractices.[39]

[39] Millennials (also referred to as Generation Y or Gen Y), are the demographic cohort born between the early 1980s and the early 2000s. Generation Z (or Gen Z) are commonly used for those born between the mid-1990s and the early 2010s. The generational boundaries are not rigid or agreed upon by all. See, for example, Michael

1.6 The Effects of Protests and Pressures

As argued in this chapter, there is much confusion about ESG—from its origins, what it is and what it is not, to its present state of evolution around the globe. There is even confusion about which part of the world it is strongest in, although there seems to be general agreement that:

> The ESG universe is still emerging in the U.S., while it has matured more in Europe and other parts of the world. We can see these signs not just from record growth rates of fund flows into ESG, but from changes in corporate policies across major companies and large money managers beginning to take a stance on the issue. (Kelly and Sardi 2021)[40]

Yes, there is consensus that ESG has made greater headway in Europe than in the United States, but evidence outside Europe, in particular, in Africa, does not show that the United States is lagging behind the developing world in terms of ESG, in spite of well-known setbacks during the Trump administration (2017–2021).

Nevertheless, one thing is certain, ESG has moved very fast from the margins to the mainstream, sweeping across the developed, and paving its way into the developing, world.[41] Both investors and issuers are under increasing pressure from multiple sources traced in this chapter. They are increasingly recognizing that their long-term successes are being linked to those of their customers, employees, suppliers, communities, and the larger societies. As Duggan and Goatman argued, "one theme will dominate headlines, markets and investment decisions in 2021. It won't be pandemics, economics or market cycles; it will be ESG".[42] 2021 was

Dimock, *Defining Generations: Where Millennials End and Generation Z Begins*. Pew Research Centre, January 17, 2019.

[40] Also refer to Emily Zulz, "ESG Bigger in Europe Than in US," ThinkAdvisor Newsletters, November 1, 2017; Andrew Holt, "ESG Investing Is More Popular in Europe Than the US," *IR Magazine*, Augest 14, 2018; Joe McGrath, "ESG's Failed U.S. Invasion," *Institutional Investor*, January 28, 2018; and Emily Holbrook, "Data Shows Broad Differences in ESG Reporting Between Europe and the US," *Environment and Energy Leader*, June 10, 2021; and David Robinson, "US Lags Europe on Regulation of ESG Investing," *Expert Investor*, June 3, 2019.

[41] Refer to LSEG (ibid.).

[42] Anthony Duggan and David Goatman, "ESG Is Set to Be the Biggest Investment Driver in Africa in 2021," Knight Frank The Intelligence Lab Global property market insight, April 7, 2021.

actually only the beginning of an acceleration of ESG globally—including Africa (as shown in Chapters 6 and 7). As this acceleration continues, the confusion about what ESG is and is not will diminish and eventually vanish as education about the topic reaches more and more people outside corporate board rooms.

1.7 New Area of Research

As we argue in this chapter, scholarly and well researched literature on formal ESG reporting in developing countries is very scarce and fragmented. On Africa, such research is almost non-existent. In fact, in our literature survey *we find no scholarly researched book on formal comprehensive or holistic ESG reporting* in developed or developing countries. However, there are many academic papers on ESG outside the continent. Additionally, there are numerous blogs, reports, books, manuals, handbooks, and academic papers on sustainability in general in the developed and developing countries.

As can be noticed in Chapter 7, most of the studies are not on Africa.[43] Our bibliography, footnotes and references in the other chapters provide additional lists of the relevant reports and studies. We find a handful of scholarly books on the related field of Corporate Social Responsibility (CSR) in Africa[44] and even fewer on Impact Investments (II) and Socially

[43] The closest book we find on comprehensive ESG reporting is John Hill, *Environmental, Social, and Governance (ESG) Investing: A Balanced Analysis of the Theory and Practice of a Sustainable Portfolio.* Academic Press, 1st edition, 2020.

[44] These are: (1) Agata Stachowicz-Stanush (ed.), *Corporate Social Performance in the Age of Irresponsibility-Cross National Perspective.* Information Age Publishing, Inc, 2017; (2) Stephen Vertigans, Samuel Idowu, and René Schmidpeter (eds.), *Corporate Social Responsibility in Sub-Saharan Africa: Sustainable Development in Its Embryonic Form.* Springer, 2016; (3) Ivan K. Mugabi, and Irene Raletsebe, "Corporate Social Responsibility in Africa: Case Study of Botswana: Botswana Companies and Global Reporting Initiative (GRI) as a Model for Sustainable Reporting," *Lap Lambert* (2017); (5) Moses Kibe Kihiko, *The African Corporate Jungle: Engaging Corporate Social Responsibility.* CreateSpace Independent Publishing Platform, 2010; (6) Nathan Andrews, and J. Andrew Grant (eds.), *Corporate Social Responsibility and Canada's Role in Africa's Extractive Sectors.* University of Toronto Press, 2020; and (7) Edward Mungai, *Impact Investing in Africa: A Guide to Sustainability for Investors, Institutions, and Entrepreneurs.* Palgrave Macmillan, 2018.

Responsible Investing in Africa. Outside South Africa, we found an odd journal paper on ESG in Nigeria.[45]

Our book offers academic and practical analyses of ESG and provides a comprehensive survey of it, including an empirical case study of South Africa in Chapter 7. The academic content includes historical, theoretical, and empirical analyses. The book is also enriched by multidisciplinary observations, analysis, and recommendations. Our research shows that it is most likely to be the first scholarly researched book on the subject with these approaches.

1.8 Structure of the Book

In what follows, Chapter 2, in the introductory part of the book outlines the particular importance of ESG in Africa. We provide an overview of the main and relevant challenges in the continent and explain the relevance of ESG in dealing with these challenges. In Part II, Chapter 3 discusses how ESG can benefit from strategic management, through the application of the African PESTLE Analysis (APA), recently advanced by the authors for Africans and non-Africans. The APA is innovatively based on the symbolism of the team of the African village woman with her pestle and mortar, and the values that the woman demonstrates which can be applied to modern strategic management. Chapter 4 discusses the COVID-19 pandemic in the context of ESG reporting priorities, far beyond "business as usual". Business as usual is defined as the pursuit of only profit by a business entity. The relevance of the APA is again shown in the chapter. Chapter 5 extends the more practical aspects of the understanding and management of ESG in order to make the book more appealing to practitioners. This is done in the chapter by introducing the processes, standards, frameworks, formats, indexes, issuers, investors, markets, and other concepts and actors necessary for the understanding, design, and management of ESG programmes. Regional case studies of ESG are carried out in Part III. The chapters in this part cover the biggest economies from each subregion of Africa. They are Egypt (North Africa), Kenya (East Africa), Nigeria (West Africa), and South Africa (Southern Africa) in Chapter 6. Due to the importance of South Africa in ESG

[45] M. Ahmed, P. F. Fashagba, and F. I. Olamide, "Effect of Environmental, Social and Governance on Corporate Performance of Listed Manufacturing's Firms in Nigeria," Federal Polytechnic Bida, Niger State, Undated.

in Africa and other reasons, we provide a separate chapter only on the country in Chapter 7. In this chapter, in addition to the common analysis in Chapter 6, we carry out an original, specific, empirical, and more detailed study of the relationships between ESG and a number of selected parameters of firm performance in a sample of listed companies on the Johannesburg Stock Exchange (JSE). Our study in the chapter is instrumental to further the quantitative understanding of the relationships between ESG and corporate performance, using South Africa (Africa's ESG pioneer and leading ESG economy) as an example. The analysis is done within the context of different theoretical perspectives.

We address the following questions in each of the countries studied in Chapter 6:

1. What are the existing national laws which directly support or can support ESG reporting?
2. What are the international treaties/agreements which the government is party to, which directly support or which can support ESG reporting?
3. Are there influential non-governmental organizations (NGOs) which directly support or can support ESG reporting?
4. Is there a stock exchange which provides for ESG reporting?

The chapters in Part III on specific countries and relevant ESG questions are intended to increase the usefulness of the book to all potential readers. These include anyone interested in economics and finance in Africa somewhat generally or in specific major economies.

The way forward for ESG disclosures/reporting, based on our findings and arguments in the book are outlined in Chapter 8. The concluding remarks in the chapter are made in relation to all key stakeholders—governments, investors, managers, workforce, NGOs, and the communities in order to enable them to effectively and efficiently pursue the ESG agenda through social contract(s).

CHAPTER 2

ESG in Africa: Relevance and Applicability

Abstract This chapter lays out the case and argues for the relevance of ESG in Africa. However, it points out that norms of international practices and conventions applied in the developed countries must not be expected overnight in Africa. This includes attention to ESG, which (as stressed and proven in this book) is a new phenomenon, about which there is little knowledge at the moment. Nevertheless, the knowledge is growing and ESG can play a very important role in tackling the continent's environmental, economic, and social challenges and, thus, enhance economic growth and benefit the private sector/businesses, communities and others in both the short and long runs. It draws attention to the possibilities that ESG offers for public–private partnerships for the national good, including the good of the businesses themselves, in Africa.

Keywords Africa · Climate change · Investment · Energy · Social protection

K. N. M. Sonko and M. Sonko, *Demystifying Environmental, Social and Governance (ESG)*, Palgrave Studies in Impact Finance, https://doi.org/10.1007/978-3-031-35867-8_2

2.1 Introduction

The ultimate objective of ESG is sustainability. According to the United Nations, despite its importance and the potential for it, sustainable development throughout Sub-Saharan Africa has consistently lagged behind those of other developing regions of the world (UNU-WIDER 2016). The concept of sustainability is relevant for companies, governments, and communities, for businesses as well as for the overall economic and social development of nations. For investors who are socially and environmentally conscious, ESG enables them to screen potential and actual investments. Such investors are guided by the relevant ESG criteria and the sense of responsible investing (RI) (Van Der Ahee & Schulschenk 2013). These investors consider both financial (corporate financial performance [CFP]) and non-financial information to guide their decision-making processes and investments (Bassen & Kovács 2008; Pasquini-Descomps et al. 2013). Thus, they come to understand the positive impacts of ESG practices on CFP (Van Der Ahee & Schulschenk 2013).

As sustainability focuses on meeting the needs of the present generations, without compromising the ability of the future generations to meet their own needs, it is not only for private or non-governmental institutions.[1] Instead, it is equally important for governments running nations and trying to develop them or maintain the levels of their development. Hence it considers the impact of current generations on the ecosystem and society (Ameer & Othman 2012). In fact, the concept of sustainability emerged in the field of economic development centuries before it did in business.[2]

Sustainability is composed of three pillars: economic, environment/ecology, and social, which are also known as the "three Ps"—People, Profit, and Planet (John Elkington 1998; Werbach 2009, 2010; Savitz

[1] In fact, we maintain throughout this book that sustainability is for all, even individuals. This book is written with the world in mind and Africa at hand.

[2] The concept was traced to Malthusian population theory in the early 1800s Justice Mensah, and Sandra Ricart Casadeval, "Sustainable Development: Meaning, History, Principles, Pillars, and Implications for Human Action: Literature Review," *Cogent Social Sciences,* Vol. 5, No. 1 (2019); J. A. Dixon, and L. A. Fallon, "The Concept of Sustainability: Origins, Extensions, and Usefulness for Policy," *Society & Natural Resources*, Vol. 2, No. 1 (1989): 73–84; J. Coomer, *Quest for a Sustainable Society*. Oxford: Pergamon, 1979.

and Weber 2013).[3] The economy, environment, and social welfare are all about people first and foremost, their businesses and the planet that they live in. Therefore, a company which considers itself sustainable, should strive to mitigate the potential negative impacts of all three pillars on people (including people on people).

The challenges faced by firms in the twenty-first century require a fundamental change in the way they function. Traditionally, firms and investors aimed to reach the "ultimate investment goal", namely to maximize return given a certain level of risk (IoDSA 2011). However, challenges such as climate change, shortcomings of fossil fuels, insufficient renewable energy availability, and the depletion of natural resources also require attention, through ESG themes, because they can have a considerable impact on the long-term sustainability of firms (Quinn & Baltes 2007; Van Der Ahee & Schulschenk 2013). A paradigm shift is therefore required in the way in which corporate managers and investors traditionally make decisions and construct their investment portfolios. This shifting paradigm, the need for it, the causes of it, and its consequences are elaborated in Chapter 1. This shift is not an easy process because it requires a high sense of conviction and level of competence that is not easy to attain.

There are various misconceptions associated with the inclusion of ESG criteria (alternatively factors, issues, themes, and various other terms in the literature.) in investment analysis and corporate decision-making and in the operations of institutions. Some investors believe that the integration of ESG would result in high-risk investment with a low return (Eccles et al. 2007). However, although reasonable thinking, this is not always necessarily the case, as Chapters 1 and 7 extensively show. Among other benefits, such as greater employee motivation when investing in a responsible manner through ESG, investors can be encouraged to act as long-term shareholders and less like short-term share traders. Active shareholders can and are expected to engage with investees on various ESG concerns (Eccles et al. 2007). The proper evaluation of a firm's ESG practices thus facilitates a better understanding of the risks and opportunities relevant to a firm (Bassen & Kovács 2008; Gond et al. 2018).

[3] For conceptual origins see Ben Purvis, Yong Mao and Darren Robinson (2019). For the triple bottom line approach see, for instance Andrew W. Savitz and Karl Weber, The Triple Bottom Line (revised and updated), Jossey-Bass 2013; John Elkington, *Cannibals with Forks: The Triple Bottom Line of the 21st Century Business*. New Society Publishers, 1998.

This could lead to improved investment decision-making. Losses associated with poor ESG performance could be avoided and the risk associated with an investment could be more accurately assessed.

In spite of its importance and growing popularity, not everyone is convinced about ESG. Conservative politicians, scholars, business persons, and social groups, again not only in Africa, are wary of its radical calls for equal opportunities for all in the work places and board rooms. Governments and companies in developing countries, not only in Africa, are generally wary about its environmental emphasis although generally they recognize the challenges that global warming is causing to the world, including Africa. Some even deny that there is anything called "global warming" or that it is caused by human activity, particularly industry. As Korstanje and George observed: "Unfortunately, the scientific community has still not reached unanimous conclusions about the causes or impacts of global warming" (2012, p. 332).[4]

In this chapter we assess the relevance of ESG to Africa. This is done by discussing each of the three categories of ESG reporting (E + S + G), one after another, in relation to the needs and realities in the continent. The potential benefits of ESG in the continent are discussed at length.

2.2 Environmental

The State of the Climate in Africa report, a multi-agency annual publication coordinated by the World Meteorological Organization (WMO), provides a credible source of the state of the climate in Africa. In this chapter, we draw upon this report and other landmark and credible studies to reflect the seriousness of the climate challenge in Africa. These official reports provide detailed information on the current and future climate conditions and trends and their consequences on the economy and society. In this chapter, we highlight lessons for climate action in Africa and identify pathways for addressing critical gaps and challenges

[4] Maximiliano E. Korstanje, "Global Warming and Tourism: Chronicles of Apocalypse?" *Worldwide Hospitality and Tourism*, Themes, Vol. 4, No. 4 (2012): 332–335; for a critical analysis of the literature see Dale S. Rothman, and Duane Chapman, "A Critical Analysis of Climate Change Policy Research," in *Global Climate Change. The Economic Costs of Mitigation and Adaptation,* ed. James C. White, William Wagner, and Carole N. Beal. Springer Link, 1991. https://link.springer.com/chapter/10.1007/978-94-011-2914-5_17.

through ESG measures. Additionally, the chapter integrates original theoretical postulates of the first author of this book 28 years ago, formally demonstrating the importance of the environment, equity and economic development. This was at a time when the world was yet to come to more seriously come to terms with the increasingly uneasy relationship between the environment and economic activity. It is opportune to recall such a historical formal and technical thinking in contemporary discourses on, and in the form of, ESG in Africa and globally. It also enriches the chapter and strengthens the case for ESG.

2.3 Climate Challenge

Africa's Agenda 2063, which was set in 2013, recognizes climate change as a major constraint to the continent's development. Since 2015, the Nationally Determined Contributions (NDCs) to favour the Paris Agreement have become the main instrument for guiding policy responses to climate change. By 2018, 52 African countries (almost all of Africa) have submitted their first NDCs and are now in the process of submitting revised NDCs.[5]

Increasing temperatures and rising sea levels, changing precipitation patterns and more extreme weather problems are threatening human health and safety, food and water security, and socio-economic development throughout Africa.[6]

2.3.1 Rising Temperatures

African temperatures in recent decades have been warming at a rate comparable to that of most of the other continents and faster than the global mean surface temperature. The year 2020 was among the four warmest years on record for Africa. This trend of rising temperatures is unfortunately expected to continue for the years to come. In fact,

[5] The revisions started in 2020. For more information, see for instance, African Development Bank (AfDB), Roadmap and Work programme: Africa NDC Hub, AfDB, Abidjan 2018a; African Development Bank (AfDB), Report: African Nationally Determined Contributions (NDCs), Abidjan 2018b.

[6] World Meteorological Organization, *State of Climate Services: Agriculture and Food Security* (WMO-No. 1242). Geneva, 2019, https://library.wmo.int/doc_num.php?explnum_id=1008.

according to the latest assessment of the Intergovernmental Panel on Climate Change (IPCC), global warming, in the form of rising temperatures and consequences such as extreme rainfall, cyclones, and droughts, is now taking place faster in Africa than in other continents, a cause of extremely serious concern given the many other challenges of Africa.[7]

The latest environmental predictions on Africa, from 2020 to 2024, show continued warming and decreasing rainfall, especially over North and Southern Africa, and increased rainfall over the Sahel. Extensive areas of Africa will exceed 2 °C of warming above pre-industrial levels by the last two decades of the twenty-first century under medium scenarios, according to the IPCC's Fifth Assessment Report.[8] On the other hand, from 1901 to 1930 Africa experienced the smallest change in temperature, rising less than 0.5 °C. All continents experienced rising mean temperature in the latest period for which actual data is available (1991–2019), with Africa having the third highest rise.[9]

2.3.2 Rising Sea Levels and Coastal Erosion

Although there are significant regional differences in sea level rises in the continent, overall rising sea levels are a growing cause of concern. Sea-level rises have reached 5 mm per year in a number of oceanic coasts and surpassed 5 mm per year in the south-western Indian Ocean areas. These areas stretch from Madagascar eastward towards and past Mauritius. This is more than the average global sea-level rise of 3–4 mm per year. Coastal degradation and erosion are also major challenges, especially in West Africa. About 56% of the coastlines in Benin, Côte d'Ivoire, Senegal, and Togo have begun to erode and this effect is expected to worsen. Although sea-level rise is presently not the main contributor, its effects, combined with those of other environmental factors, are expected

[7] Kiran Pandey, "Africa Warming Faster than Rest of World: IPCC Sixth Assessment Report," *DownToEarth*, March 29, 2022, https://www.downtoearth.org.in/.

[8] IPCC, "Climate Change 2014: Synthesis Report," in *Contribution of Working Groups I, II and III to the Fifth Assessment Report of the Intergovernmental Panel on Climate Change*, ed. Core Writing Team, R. K. Pachauri, and L. A. Meyer. Geneva, Switzerland: IPCC, 2014; for a summary of the IPCCC Report, the interested reader may also see Alan Buis, "A Degree of Concern: Why Global Temperatures Matter," *NASA*, June 19, 2019, https://climate.nasa.gov/news/2865/a-degree-of-concern-why-global-temperatures-matter/, March 29, 2022.

[9] WMO, *State of the Climate in Africa 2019*. Geneva: WMO, 2020, Fig. 2, p. 6.

to exacerbate the negative consequences of climate change in Africa in the future. Some African cities, such as Banjul, the capital of the Gambia, are under the threat of being completely engulfed by the sea.[10]

2.3.3 *Extreme Events*

The 2020 WMO report documents high-impact events. Tropical Cyclone Idai was among the most destructive tropical cyclones ever recorded in the southern hemisphere, causing hundreds of casualties and displacing hundreds of thousands of victims. Southern Africa suffered extensive drought in 2019. In contrast, the Greater Horn of Africa moved from very dry conditions in 2018 and for most of 2019 to floods and landslides with heavy rainfall in late 2019. Flooding also affected the Sahel and surrounding areas from May to October 2019.

In 2020, flooding took place in different parts of Africa, with Kenya and Sudan suffering the worst consequences. There were 285 reported deaths in Kenya, and 155 deaths and more than 800,000 people displaced or affected in other ways in the Sudan. There were further indirect impacts, including diseases. Other countries which suffered from losses of life or significant displacement of people included Cameroon, Chad, Ethiopia, Somalia, South Sudan, Nigeria (whilst drought was nothing place in the Southern part), Niger, Uganda, Benin, Togo, Senegal, Côte d'Ivoire, and Burkina Faso. Lake Victoria, the Niger River at Niamey, the Blue Nile at Khartoum, and many other rivers and lakes rose to unusually high levels. In contrast, long-term severe drought persisted in the Northern and Eastern Cape Provinces of South Africa and other parts of Southern Africa.

On 22 November 2020, Tropical Cyclone Gati, originating from the Bay of Bengal, became the strongest storm ever to hit Somalia, pouring more than a year's worth of rain in only two days.[11] According to the WMO, this was the first cyclone to make landfall in Somalia as a category 2 storm on the Saffir–Simpson scale. The storm brought heavy rain to

[10] The State of the Climate in Africa, 2019 and 2020 reports, World Meteorological Organization (WMO), Geneva 2019 and 2020.

[11] UN Migration Agency, "Climate Change Drives Displacement of Thousands of Vulnerable People in Somalia," January 4, 2021, accessible at https://medium.com/@UNmigration/climate-change-drives-displacement-of-thousands-of-vulnerable-people-in-somalia-166171bb52c3, last visited September 18, 2022.

the region, and local authorities reported that at least nine people were killed and tens of thousands displaced. Thousands of properties, including 10,000 livestock, belonging to nomadic communities, were destroyed.

2.3.4 *Food Security Impacts*

Agriculture is the backbone of African economies and accounts for the livelihoods of the majority of the people across the continent. African agriculture is affected by climate variability and change impacts. Key risks to agriculture include reduced crop productivity associated with heat and drought and increased post-harvest damages, pest and disease damages, and flood impacts on food systems infrastructure. These in turn have adverse consequences on food security and livelihoods. The number of undernourished people in Sub-Saharan African countries has increased by 45.6% due to droughts since 2012, according to the Food and Agriculture Organization of the United Nations (FAO).[12]

By the middle of this century, major cereal crops grown in Africa will be adversely affected by climate change, albeit with regional variability and differences according to the crop. The UN's forecasts for cereals, under the worst case scenario, show a reduction in mean yield of 13% in West and Central Africa, 11% in North Africa, and 8% in East and Southern Africa. Millet and sorghum have been found to be the most resistant crops, with estimated yield losses of 5% and 8%, respectively, by 2050, because of their greater resilience to heat-stress. Rice and wheat are expected to be the most affected crops with yield losses of 12% and 21%, respectively, by 2050.[13] The cases of these two (rice and wheat) are particularly worrying because rice is the stable diet in many African countries and wheat consumption in the continent is growing rapidly.[14]

[12] The State of the Climate in Africa 2019 report, World Meteorological Organization (WMO); FAO, "State of Food Insecurity in the World In Brief," 2014, https://www.fao.org/3/i4037e/i4037e.pdf, July 15, 2022; FAO et al., *The State of Food Security and Nutrition in the World. Safeguarding Against Economic Slowdowns and Downturns*. FAO, 2019 and 2021.

[13] UN Climate Change, *Climate Change Is an Increasing Threat to Africa*. Bonn, Germany: UNCC, October 27, 2020. Also see Dimitiri Defrance et al., "Impact of Climate Change in West Africa on Cereal Production Per Capita in 2050," *Sustainability*, Vol. 12 (2020): 7585, https://doi.org/10.3390/su12187585.

[14] The consumption of rice and wheat have expanded in Sub-Africa in per capita terms. In West Africa, for instance, it is now more than 60 kilogrammes per capita annually,

2.3.5 *Health Impacts*

Increases in temperature and changes in rainfall patterns significantly affect people's health too. Warmer temperatures and higher rainfall create ideal conditions for biting insects and the transmission of vector-borne diseases, such as dengue fever, malaria, and yellow fever. In addition, new diseases are emerging in regions where they were previously not present. In 2017, an estimated 93% of global malaria deaths occurred in Africa. Malaria epidemics often occur after periods of unusually heavy rainfall. In addition, warming in the East African highlands is allowing mosquitoes to survive at higher altitudes, unlike the cooler temperatures of the past which were inhospitable to mosquitoes. No wonder, in many parts of Africa people are now experiencing diseases not known before or known but with lesser occurrences than now.

2.3.6 *Economic Impacts*

According to the International Monetary Fund (IMF), the most adverse consequences of climate change are concentrated in regions with relatively hot climates, where a disproportionately large number of low-income countries are, rather unfortunately, also located.[15] These regions include all of Africa, especially the Northern part which is mostly desert. About 12% of population displacements worldwide in 2020 happened in the East and Horn of Africa, where more than 1.2 million displacements and about 500,000 conflict-related human displacements took place. The biggest contributors were floods and storms, then droughts. The costs of adaptation in Sub-Saharan Africa are estimated to range between US$ 30 and 50 billion (2–3% of regional GDP) per annum, without taking into account the costs of potential new disasters.

on average, and more than the global average. Getachew Nigatu, James Hansen, Nathan Childs, and Ralph Seeley, "Sub-Saharan Africa Is Projected To Be the Leader in Global Rice Imports," in *USDA Agricultural Projections to 2026*, ed. Erik O'Donoghue, and James Hansen, USDA Agricultural Projections to 2026. OCE, February 2017. Wheat consumption, as shown by imports, surged by 68% from 2007 to 2019; Michael King, "Wheat Consumption Rising in Africa," *World.Grain.com*, https://www.world-grain.com/articles/11655-wheat-consumption-rising-in-africa, July 15, 2022.

[15] The State of the Climate in Africa 2019 report, World Meteorological Organization (WMO).

Table 2.1 Impacts of climate change on Africa's GDP (%) according to 4 global scenarios

Subregions	*GDP (% Change/Year)*			
	1 °C	*2 °C*	*3 °C*	*4 °C*
North ($n = 7$)	−0.76 ± 0.16	−1.63 ± 0.36	−2.72 ± 0.61	−4.11 ± 0.97
West ($n = 15$)	−4.46 ± 0.63	−9.79 ± 1.35	−15.62 ± 2.08	−22.09 ± 2.78
Central ($n = 9$)	−1.17 ± 0.45	−2.82 ± 1.10	−5.53 ± 1.56	−9.13 ± 2.16
East ($n = 14$)	−2.01 ± 0.20	−4.51 ± 0.34	−7.55 ± 0.63	−11.16 ± 0.85
Southern ($n = 10$)	−1.18 ± 0.64	−2.68 ± 1.54	−4.40 ± 2.56	−6.49 ± 3.75
Whole of Africa ($n = 55$)	−2.25 ± 1.52	−5.01 ± 3.30	−8.28 ± 5.12	−12.12 ± 7.04

Source Adapted from Economic growth, development, and climate change in Africa, published by the African Climate Policy Centre (ACPC) of the United Nations Economic Commission for Africa (UNECA)

The African Climate Policy Centre projects that the Gross Domestic Products in the five African subregions of East, West, South, North, and Central Africa would suffer significant decreases as a result of global temperature rises. For scenarios ranging from a 1 to a 4 °C increase in global temperatures, relative to pre-industrial levels, the continent's overall GDP is expected to decrease by 2.25–12.12% (WMO 2019). West, Central, and East Africa exhibit a higher adverse impact than Southern and North Africa (see Table 2.1).

Based on the evidence so far in this chapter, it has become clear that climate-resilient development in Africa calls for ESG in the private sector, in addition to a more active and efficient role of the state in dealing with high-impact natural disasters. Next, in addition to this evidence, we also go back to an original theoretical justification of the authors, demonstrating the importance of the environment in ESG in Africa and globally.

2.3.7 *The Environment and Equity*

Khan and (the first author of this book) Sonko (1994) have formally made a contribution to the pioneering theoretical development of arguments of the linkages between the environment and economic development almost thirty years ago, demonstrating the relationship through income inequality and poverty.[16] We referred to this as the "eco-economic" approach, which also incorporates sustainability as the need to create an equilibrium or a balance between the needs of the present and future generations (Fig. 2.1).[17] Our work (1994 and 1995, part of which is presented below with reference to ESG) (was motivated by the observation even then that "It has become clear now that the ecology imposes clear limits to economic activity which if overstepped could have long term disastrous consequences" (1994, p. 189). Based on the evidence in this chapter, this observation and our postulates next are more relevant today than in 1994 and our technical analysis provide very strong justification for arguments in favour of ESG reporting in Africa. The theory springs from and focuses on the environment degradation, the consequences of which on income inequality and poverty lead to the "social" (directly) and "governance" (indirectly) in ESG. This is because the environment is linked to production, distribution, inequality and poverty. As externalities distort income, we take this into account by redefining true income as income as total income adjusted for environmental effects, which ESG programmes aim to deal with. Thus, we also show that the real level of inequality and poverty is most likely higher than the commonly measured levels, which again calls for social attention (the "S" in ESG).[18] Clearly, the analysis below applies development, ecological, and mathematical economics to a new business concept—comprehensive ESG.

Our analysis begins by developing and using the definitions, axioms, and theorems of income inequality comparisons. This axiomatic approach to inequality comparisons usually utilizes three axioms. Of these, the axiom of symmetry states that distribution is not individual-specific; in

[16] Haider A. Khan, and Karamo Sonko, "A Further Extension of Adjustment Models,' in *Economic Justice in Africa: Adjustment and Sustainable Development in Africa*, ed. George W. Shepherd, and Karamo Sonko. Westport, CT: Praeger Publishers Inc., 1994.

[17] In line with Norgaard's observation with regard to sustainable development in economics, Fig. 2.1 shows that there are multiple efficient shows multiple solutions depending on the initial state of the distribution of rights or entitlements. For example, R. B. Norgaard, "Sustainability and the Economics of Assuming Assets for Future Generations," World Bank Staff Working Papers, No. WPS832, 1992.

[18] The approach is developed by Khan and Parvin (1984).

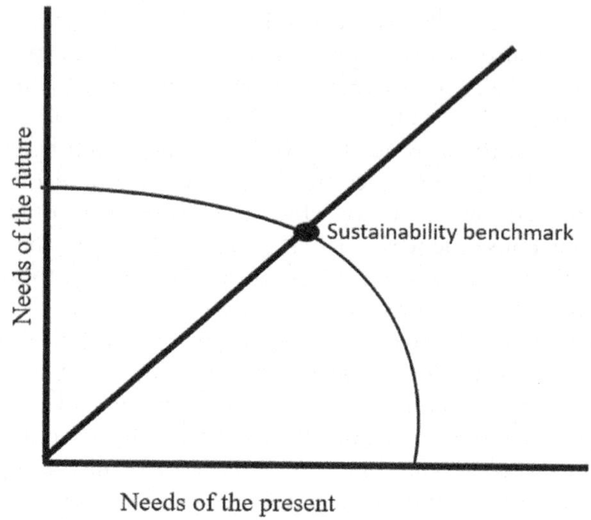

Fig. 2.1 Sustainability (*Source* Khan and Sonko, 1994)

other words, interchanging incomes of two individuals has no effect on inequality.[19]

2.3.7.1 Axioms for Inequality Comparisons
Let

$$X = (X_1, X_2, \ldots, X_i, \ldots, X_n)$$
$$\left(X_i \epsilon R, n < \infty, \text{and } X_i < X_j \text{ if and only if } i < j\right) \quad (2.1)$$

Represents income distribution of n households (individuals). Each X_i is a scalar, and X is an element of R^n, the n-dimensional Euclidean space excluding the origin (0, ..., 0). We allow true income to be negative though observed income is always negative as in Fields and Fei (1978).[20]

[19] In an axiomatic approach to inequality comparisons, Fields and Fei introduce the three axioms in Gary S. Fields, and John C. H. Fei, "On Inequality Comparisons", *Econometrica*, Vol. 46, No. 2 (March, 1978).

[20] For mathematical convenience we use R^n. It should be noted that income can be expressed by integers through the smallest monetary unit available (e.g. counting by pennies). Accordingly, the rational numbers will suffice for normalization of income. For

Income here is net of taxes. Following them, we now introduce the method of comparison of income distributions. Given any two income distributions X and Y, we would like to know if.

$$\text{(i)}\quad X \text{ is more equal than } Y \quad (X > Y) \tag{2.2a}$$

$$\text{(ii)}\quad Y \text{ is more equal than } X \quad (Y > X)$$

$$\text{(iii)}\quad X \text{ and } Y \text{ are equally unequal} \quad (X \approx Y)$$

The binary relations $\geq$ on $R_{,}^{n}$, reading "is at least as equal as," or "is not unequal than," satisfy the following preordering conditions.[21]

Comparability: $\forall X, Y \in R_{,}^{n}$, exactly one of the following must be true:

$$\text{(i)}\quad X > Y \tag{2.2b}$$

$$\text{(ii)}\quad Y > X \tag{2.2c}$$

$$\text{(iii)}\quad X \geq Y \text{ and } Y \geq X \text{ in which case } X \approx Y \text{ to be read as "}X \text{ and } Y \text{ are equally unequal."}$$

$$\text{Reflexivity: } X \geq Y$$

$$\text{Transitivity: } X \geq Y \text{ and } Y \geq Y \geq Z \text{ implies } X \geq Z. \tag{2.2d}$$

Obviously, the relation > will not satisfy (2.2c). Note that reflexivity is not required for completeness.

Symbolically,

Axiom 1 Axiom of Symmetry[22]

$$\text{If } X' = \left(X_1', X_2', \ldots, X_n'\right) \text{is obtained from}$$

cardinal analysis, Fields and Fei include irrational numbers in X and exclude negative numbers, relative to us.

[21] Debreu (1959) defines a reordering as a relation that is reflexive and transitive. Fields and Fei do not include reflexivity.

[22] A. K. Sen, *On Economic Inequality*. Oxford: Clarendon Press, 1973, calls this expressively the axiom of anonymity.

$X = (X_1, X_2, \ldots, X_n)$, where X and X' are income distributions.

By a permutation of indices $(1, 2, \ldots, n)$,

Then $X' \approx X$.

Axion 2 Axiom of Rank-preserving Equalization (RPE)

This axiom states that if income is redistributed from higher- to lower-income households such that no ranks are changed (i.e. if $i < j$ then $X_i < X_j$ for all i and j in both distributions), then the new distribution is more equal.

To put it formally,

Definition: An income distribution X resulting from a progressive redistribution of Υ is rank preserving $[X = E(\Upsilon)]$ if and only for some i,j $(i < j)$ and $h > 0$

$$\begin{aligned} X_k &= Y_k \text{ for } k \neq i, j; \\ X_i &= Y_i + h; \\ X_j &= Y_j - h; \end{aligned} \tag{2.3a}$$

where

$$\begin{aligned} &\text{if } j = 1 + 1, h \leq \frac{1}{2}(Y_j - Y_i); \\ &\text{if } j \leq i + 1, h \leq \min\left[(Y_{i+1} - Y_i), (Y_j - Y_{j-i})\right], \end{aligned} \tag{2.3b}$$

RPE tells us that if $X = E(\Upsilon)$, then $X > \Upsilon$.[23]

Finally, Axiom 3 states that if incomes are proportionally increased (or decreased), then inequality remains the same.

To put it formally, Axiom 3. Axiom of Scale Irrelevance.[24]

$$\text{if } X = aY \ (a > 0), \text{ then } X \simeq Y.$$

[23] Dalton terms Axiom 2 the "Principle of Transfers." H. Dalton, "The Measurement of the Inequality of Incomes," *The Economic Journal*, Vol. 30 (September 1920): 348–351.

[24] The axiom of scale irrelevance is not as reasonable as the other two axioms. If material well-being is the ultimate object of inquiry, then the absolute *level* of income cannot be ignored. However, if the axiom relates to units of measurement, then this objection does not arise.

This marks the comparison of inequality a matter of relative income comparisons and not of absolute levels of income.

The above axioms are used to derive the following fundamental theorem, which in turn is applied to prove the propositions required for ordinal inequality comparisons.[25] If individuals with lower income lose a greater portion of their income than individuals with a higher income, then income distribution becomes less equal (or more unequal).[26]

2.3.7.2 *External Costs and Income Inequality Comparisons*

These axioms can be used to investigate the implications of the inclusion of external costs of the environment for income inequality comparisons. Initially, the impact of government regulations on income distribution is ignored and the earnings loss caused by externalities is excluded, the reason for the sequential approach is that it distinguishes the economic results under different conditions before general results on true income distributions are presented.

2.3.7.3 *Adjusted Income*

Let us define adjusted income as the difference between observed income and the external cost. Let h_i be the out-of-pocket health cost attributed to externalities.[27]

Even though h_i is greater for individuals with lower income, we can, for the sake of simplicity, assume a uniform distribution of hi without loss of generality.[28]

So that

$$h_i = h > 0 \tag{2.4a}$$

[25] The three axioms are not sufficient for inequality comparisons in all cases. Some comparisons are ambiguous. See pp. 308 and 312–313 of Gary S. Fields, and John C. H. Fei, "On Inequality Comparisons," *Econometrica*, Vol. 46, No. 10 (March 1978): 303–316.

[26] Khan and Parvin, "Environment, Externalities and Income."

[27] Note that at this stage not all opportunity costs (e.g. earning loss due to illness) are included and all public goods are excluded.

[28] It is known that families living in inner-city slums are subjected to a proportionally greater amount of negative externalities like crime, pollution, and noise.

Adjusted income for i, X_i^h, is defined by

$$X_i^h = X_i - h_i = X_i - h$$

Note that if $h_i > X_i$, then the adjusted income is $X_i - h_i < 0$.

For analytical purposes, X_i^h is transformed into a fraction of observed income:

$$X_i^h = a_i x_i a_i < 1. \quad (2.4b)$$

In order to use the fundamental theorem to compare the true with observed income distribution, the following lemma is proved.

Lemma 1 If h_i is a non-increasing sequence,[29] then a_i is an increasing sequence

Proof

$$X_i^h = X_i - h_i = a_i X_i,$$

$$a_i = 1 - \frac{h_i}{X_i}.$$

According to our definition $X_i > X_j$ if and only if $i > j$;

$$\forall i, j \text{ such that } i > j,$$

$$X_i > X_j, \text{and } \frac{h_i}{X_i} > \frac{h_j}{X_i}.$$

So,

$$a_i = 1 - \frac{h_i}{X_i} > 1 - \frac{h_j}{X_i} = a_j.$$

Note that Lemma 1 holds in particular when $h_i = h =$ constant.

This allows us to establish the following propositions about inequality comparisons.

[29] It is possible to construct an increasing h_i that could not violate our propositions, but that is not necessary.

Proposition 1 If $h_i = h > 0$, then $X > X^h$.

Proof

Suppose $X^h = (a_1X_1, a_2X_2, \ldots, a_nX_n)$.

By Lemma 1, $a_1 < a_2 \ldots < a_n < 1$.

By the fundamental theorem $X > X^h$.

The assumption above of uniform distribution of external cost has caused income at the lower end of the distribution to decrease proportionately more than that at the upper end. This fact enables us to use rank-preserving equalization (Axiom 2) repeatedly to show that even with egalitarian distribution of bads the effect of external cost increases inequality.

Proposition 1 leads logically to Corollary 1:

Corollary 1 If $\{h_i\}_{i=1}^{h}$ is a decreasing sequence, then $X > X^h$.

By the above assumption, the true income at the low end of the distribution decreases proportionately more than that at the upper end. Hence the income distribution is more unequal when negative externalities are included in the definition of income.

2.3.7.4 *Earnings Loss and Inequality*

If externality-induced loss of earnings as a result of the effects of the environment on human health (stated earlier in this chapter) is included in the definition of income, then it becomes possible to compare the resulting income distribution.

For simplicity, assume $h_i = h \geq 0$ and assume also that the time lost, t_i because of environmentally related illness is equal for all $i, t_i = t =$ constant. Assume further that individual in the lower end of income is paid by piece or time work, whilst individual j in the higher level receives salaries, interest, and rents. It follows that the portion of income lost by I because of illness is greater than that of j, since not all income of j is time related, $t_{j>i}$. We can now show that potential income distribution, X^P, (in the absence of enviromentally related illness) would have been more equally distributed than observed income, X_i.

Where.

$$X_i^p = X_i + e_i = X_i + b_iX_ib_i \geq 0,$$

And e_i is earnings loss due to external effect.

Proposition 2 If $\{b_i\}_{i=1}^n$ is a decreasing sequence, then $X^P > X$.
Proof

$$X_i^P = X_i + e_i = X_i + b_i X_i = (1 + b_i)X_i, \tag{2.5a}$$

$$\text{By hypothesis, } b_1 > b_2 > \ldots > b_n. \tag{2.5b}$$

$$\text{Therefore, } \frac{1}{1+b_1} < \frac{1}{1+b_2} < \ldots < \frac{1}{1+b_n}. \tag{2.5c}$$

From (2.5a), $X_i = \frac{1}{1+b_1} X_i^P = a_i X_i^P$, where the proportion of X_i^P which is observed is $a_i = \frac{1}{1+b_1}$.

Then by the fundamental theorem

$$X^P > X.$$

Therefore, when external effects impose a greater proportional loss on potential income at lower-income brackets, then the observed income distribution is even more unequal than the one without externalities.

It is obvious that the above results would be greater if we assumed that time loss due to environmentally related illness is more at lower-income brackets than at higher brackets, that is,[30]

$$t_i > t.$$

The concept of X^P in contrast to X reveals the inexactness introduced in inequality comparisons by not considering all effects of external costs on income distribution.

2.3.7.5 *External Cost and Poverty*

The global environmental "crisis" has added to the reality and threats of poverty as a universal phenomenon. It has also brought into focus the necessity to consider all factors that affect true poverty versus reported poverty indicated by observed income, so that the extent and nature of true poverty will not be significantly underestimated.

[30] See, for example, Daniel Hammermesh, and Albert Rees, *Economics of Work and Pay*. New York: Harper & Row, 1984, p. 285.

Next, we highlight ignored dimensions of poverty that deserve attention for more effective anti-poverty measures which are or should be inherent in ESG in Africa and globally.

Definition 1 A poverty line is an income level, $L > 0$.

Definition 2 The Headcount (H) refers to the number of individuals whose income is below L (also called the number of the poor).

Definition 3 Shortfall (SF) is the sum of the differences between the poverty line and the incomes of the poor, that is,

$$SF = \sum_{i=1}^{k} (L - X_i),$$

where $X_k \leq L > X_{k+1}$. Therefore, $H = k$.

Definition 4 Per capital (or average) shortfall $S = \frac{1}{k} SF$.

Definition 5 Poverty surplus (PS) is the sum of the differences between the income of the non-poor and the poverty line.

$$PS = \sum_{i=k+1}^{n} (X_i - L),$$

Which is a mirror image of SF. Assuming a uniform distribution of bads, we can prove the following:

Proposition 3 Inclusion of the external cost increases H if $h > |L - X_{k+1}|$; otherwise, H does not change.

Proof

$$X^h = \left(X_1^h, X_2^h, \ldots, X_k^h, \ldots, X_n^h\right).$$

$$\text{If } h > |L - X_{k+1}|, \text{ then } X_{k+1}^h = X_{k+1} - h \leq L.$$

Hence, H increases by at least one individual.

Proposition 4 External cost increases total shortfall and decreases the poverty surplus.

Proof

Shortfall, with external cost,

$$SF^h = \sum_{i=1}^{k'} (L - X_i^h), \tag{2.6}$$

where $k' \geq k$.

Equation (2.6) can be expanded to

$$SF^h = \sum_{i=1}^{k} (L - X_i^h) + kh + \sum_{i=k+1}^{k'} (L - X_i^h).$$

Since

$$kh > 0 \text{ and } \sum_{i=k+1}^{k'} (L - X_i) = SF => 0,$$

Then

$$SF^n = \sum_{i=1}^{k} (L - X_i) = SF.$$

It is obvious that the per capita shortfall for the original poor will also increase, that is,

$$\frac{SFh}{k} > \frac{SF}{k}.$$

Note that per capita *SF* of the k and the SF^h of the k' are not comparable unless $k' = k$. The statement

$$\frac{SF}{k} < \frac{SF^h}{k'}$$

is not valid in all cases. But it is clear that external cost has left everyone worse off.

The poverty surplus including external costs,

$$PS^h = \sum_{i=k\prime}^{n}(X_i^h - L)$$

$$= \sum_{i=k\prime}^{n}(X_i - L) - (n - k\prime)h$$

$$= \sum_{i=k'}^{n}(X_i - L) \text{ since } (n - k')h > 0.$$

Therefore $PS^h < PS$.

Our analysis shows that the external costs worsen poverty, measured by *SF* and *H* in general, as well as by other indices as well.[31] Our theoretical demonstrations here provide further justification for taking the environmental impact on equity and poverty into consideration in business in order to attain sustainability. Khan and Sonko (1991) have made a call for this in the designs of the International Monetary Fund (IMF) and World Bank's controversial Structural Adjustment Programmes (SAPs) at the time and Khan (1991) also made so in the disbursement of Japanese foreign aid.[32]

2.4 Social

The "Social" in ESG in Africa is as important as the environment. In fact, most people may argue it is more important because it is directly about people. People must be at the centre of ESG reporting because neither environmental protection nor good corporate governance are meaningful unless they are beneficial to human beings. Therefore, the environment can be the beginning of ESG and is extremely important but must not be an end in itself. In the preceding paragraphs of Sect. 2.3.2 (above), we highlighted Africa's climate threats, if not crisis, as a way of demonstrating the importance of the "E" in ESG in Africa. Next, we discuss the

[31] The case of the Sen Index (1974), for instance, has been discussed in Khan and Parvin, "Environment, Externalities and Income," unpublished paper, University of Denver, 1990.

[32] See H. A. Khan, "Japanese Foreign Aid," *Journal fur Entwicklungs Politik* (1994).

reasons why the "Social" in ESG is also relevant to Africa. This is done by focusing on the importance of corporate involvement in addressing poverty in general through shared prosperity. Then we address the need to specifically pay attention to education, health, empowerment of the three most vulnerable groups in African society (women, youth and children), the rural areas, MSMEs, and to the essential process of addressing the needs of everyone through social protection.

The recommendations here are for voluntary corporate social responsibility in the specified areas. Companies are part of the socio-political and natural environments. They owe society and have responsibilities to it. However, they are not governments and they cannot replace governments. Their primary obligations to society are to pay taxes and create employment and behave responsibly in these processes. Expecting them to take over the roles of governments in proving public infrastructure and services and directly addressing the fundamental causes of economic and political instability in Africa is not realistic and can in fact result in ineffective state governance. Therefore, our position here is that corporate social responsibility, as part of ESG, must be secondary to the role of governments in economic development. It must be supplementary, voluntary, mainly out of self-interest, and that the main contributions of companies in economic development must be through statutory obligations out of their revenues and paid to government coffers and properly utilized by government officials in the interest of their citizens.

2.4.1 The Women

More than 50% of Africa's population are women, but they produce only 33% of the continent's combined GDP, according to Mckinsey Global Institute (2019). In the same study, Mckinsey (ibid.) also found that women's education and financial inclusion in Africa compared to those of men are less than the world averages, although there has been a general decline in financial inclusion of women globally, not only in Africa.[33] They observed further that violence against women is also a global issue, but that it is worse in Africa than the world average. These concerns create an opportunity for locally owned businesses and foreign businesses operating

[33] Mckinsey Global Institute, "COVID-19 and Gender Equity: Countering the Regressive Effects," July 15, 2020; Mckinsey Global Institute, "Women in the Workplace 2021," September 27, 2021.

in Africa to engage in programmes which economically empower women. They can do so directly, through NGOs and/or through international development finance institutions active in providing support to MSMEs, where most female entrepreneurs are categorized. One of the examples of such an institution is the International Financial Corporation (IFC) of the World Bank. The Corporation estimated in one of its studies on African SMEs that the credit gap for women-owned SMEs is between $26 and −320 billion, although women's repayment record throughout Africa is generally better than that of men.[34] To deal with the credit shortfall, the IFC launched the Banking on Women for women-owned SMEs in 2010. The Corporation also provides non-financial assistance to MSMEs, such as advisory services.

2.4.2 The Youth

The total African population has now reached about 1.3 billion and growing faster than that of any other continent. It has been projected to double by 2050.[35] The median age in 2020 was 19.7 and most of the population are under the age of 35—in other words, the youth (Table 2.2).[36] In 2020, this age cohort amounted to almost a billion people in all of Africa and about 22.7% of the world's youth population. This population of the youth has been projected to reach 47.7% in 2100, overtaking

[34] For Instance, in 2015, 36 IFC-supported financial intermediaries in Sub-Saharan Africa provided gender-disaggregated data which showed that the non-performing loans of the women borrowers were lower than the NPLs or those of the men. Saulo Teodoro Ferreira, *Sub-Saharan Africa: IFC Financing to Micro, Small and Medium Enterprises*. Washington, DC: IFC, n.d., p. 2, https://www.smefinanceforum.org/sites/default/files/IFC_Factsheets_SSA.pdf, March 28, 2022.

[35] United Nations, "Global Issues. Population," https://www.un.org/en/global-issues/population, last visited December 4, 2021; https://www.pewforum.org/2011/01/27/future-of-the-global-muslim-population-regional-sub-saharan-africa/regional-sub-saharan-africa/, last visited December 2, 2021.

[36] The definition of the youth is not standard. Although always defined according to the age, the age bracket or the lower and upper limits for what is defined as the youth may be different according to the author or organization. The most official definition is that by the UN "those persons between the ages of 15 and 24 years," endorsed by the UN General Assembly in resolution 36/28 of 1981. See UN, Global Issues. Youth. https://www.un.org/en/global-issues/youth#:~:text=Today%2C%20there%20are%201.2%20billion,cent%2C%20to%20nearly%201.3%20billion.

Table 2.2 Regional median age comparisons in 2020 and 2100

Africa	*2020*	*2100*
Africa	19.7	34.4
Asia	32.0	46.7
Europe	42.5	48.2
Latin America & Caribbean	31.0	49.3
Northern America	33.4	42.6
Oceania	38.4	45.6

Source Authors, data from Rocca and Schultes (2021)

Asia's share of 37.8% (the second highest).[37] By 2100, Africa's youth will account for about 50% of the total youth population of the world and be equivalent to twice the entire population of Europe.[38]

In Sub-Saharan Africa, children and young persons between the ages of 0–34 years account for almost 80% of the poor. As this population increases and the numbers of young persons and women increase too, concerns about their welfare and, hence, financial inclusion, grow simultaneously. These concerns create an opportunity for locally owned and foreign businesses operating in Africa to engage in programmes which economically empower the youth; in other words the "social" in ESG. They can do so directly, through NGOs or with international development finance institutions.

2.4.3 *The Children*

Worldwide, women and children are the greatest victims of violence and poverty. With regard to poverty, children below the age of 15 account for 50% of the world's poor, although they make up for 25% of the total population. The situation is worse in Sub-Saharan Africa, where children from the ages of 0–14 account for about 50% of those below the international poverty line of $1.90. In South Africa, for example, as many as two-thirds of children are living in poverty, with 2.1 million of them still

[37] Camilla Rocca, and Ines Schultes, "Africa's Youth: Action Needed Now to Support the Continent's Greatest Asset," *Mo Ibrahim Foundation*, available at https://mo.ibrahim.foundation/sites/default/files/2020-08/international-youth-day-research-brief.pdf08/international-youth-day-research-brief.pdf, last visited 12 December 2021.

[38] Rocca and Schultes (ibid.).

unable to access the Child Support Grant, which provides assistance to children and has so far benefitted at least 10 million (UNICEF, n.d.). COVID-19 has worsened this situation, as the World Bank observed through its Human Capital Index (HCI), which measures children's future productivity relative to a benchmark of good health and complete education: "Before COVID-19, for example, a child born in Sub-Saharan Africa could expect to achieve only 40% of her potential productivity as an adult worker, given shortfalls in health and education in the region, meaning that an African child would grow up to be just 58% as productive as a child raised in Europe and Central Asia".[39] ESG programmes are therefore important for tapping the potential of Africa's growing youth population, who happen to be also the source of present and future labour supplies to the public and private sectors.

2.4.4 *Education*

There are a million reasons why education is essential in economic development and in almost everything else in life! That is why it has become a basic human right. Yet it remains elusive and a privilege beyond a certain level in many African countries. This level is normally restricted to primary school, which is now universal or free in most African countries.[40] However, many Africans, for one reason or another, do not attain even this level. In Sub-Saharan African more than 30% of the population below the poverty line have no education at all and more than 90% have only up to secondary level. The adult literacy and primary enrolment rates are shown in Fig. 2.2.

2.4.5 *Health*

Access to health care is the most critical challenge for social protection in Africa. Healthcare services beyond the primary levels are rare, underdeveloped and too expensive for most of the populations. In Kenya and Senegal, for example, 45% of total health expenditure is paid personally

[39] World Bank, *Global Economic Prospects*. Washington, DC: World Bank, June 2020.

[40] 42 African countries have free primary school education (UNESCO and BBC World Service, "Can Africa Afford Free Education?" January 27, 2016) (updated on: February 2, 2022). http://uis.unesco.org/sites/default/files/documents/universal-primary-education-in-africa-the-teacher-challenge-en.pdf.

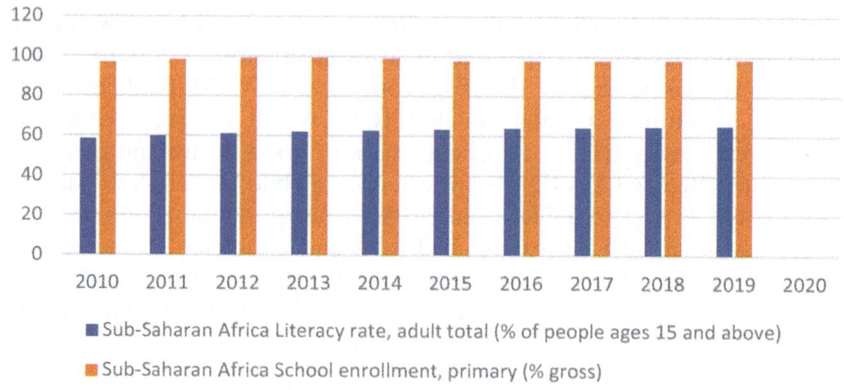

Fig. 2.2 Sub-Saharan Africa adult literacy and primary school enrollment rates 2010–2019 (*Source* Authors, based on data from various sources)

by patients. Access to clean water and proper sanitation are also elusive for most Africans (see Fig. 2.3). The consequences are the prevalence of diseases and relatively low life expectancies at birth, as shown in Fig. 2.4.

Fig. 2.3 Sub-Saharan Africa people using safely managed water and sanitation services (*Source* Authors, based on data from various sources)

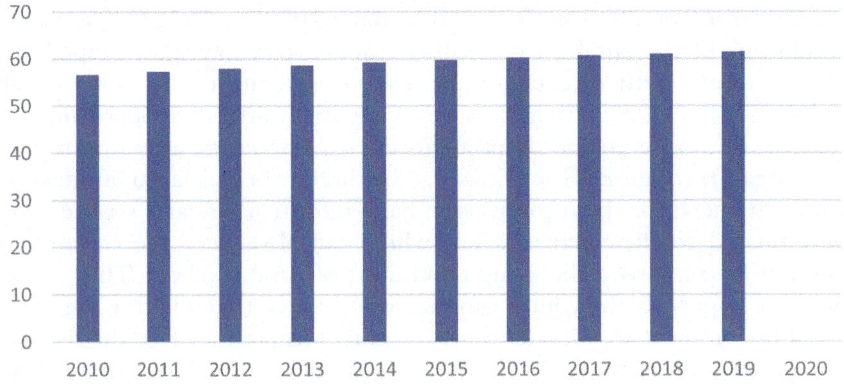

Fig. 2.4 Sub-Saharan Africa Life expectancy at birth, total (years) (*Source* Authors, based on data from various sources)

2.4.6 The Rural Areas

Evidence shows that the percentage of the poor, worldwide, is increasing. It also shows that four out of every five persons below the poverty line are from the rural areas, although the rural population accounts for about 48% of the world's total population (World Bank (2020).[41] In spite of massive rural–urban migration over the years, 59% of Sub-Saharan Africans still live in the rural areas.[42] In these areas throughout Africa, modern infrastructure and services and opportunities for formal and properly paid employment are rarest. Poverty is at its most acute. "In such settings, economic returns to people's work and skills are low, assets slowly accumulate but can quickly erode, vulnerability to risks is high, social and physical connections to markets are few, discrimination is common, people's ability to influence broader policy decisions is weak, and public services are of low quality. The consequences for human capital

[41] World Bank. *Poverty and Shared Prosperity 2020: Reversals of Fortune*. Washington, DC: World Bank. https://doi.org/10.1596/978-1-4648-1602-4.

[42] World Bank, *Rural Population (% of Total Population)—Sub-Saharan Africa 2020*. World Development Indicators. https://data.worldbank.org/indicator/SP.RUR.TOTL.ZS?locations=ZG, last visited March 20, 2022.

accumulation are often severe" (World Bank 2020, p. 163).[43] The harsh conditions of the rural areas result in rural–urban migration, especially of the youth, with dire consequences on agriculture and other rural economic sectors, including areas were potentially attractive opportunities exist for both foreign and domestic investors. Therefore, attention to the rural areas, through ESG activities, can be directly beneficial to businesses; hence the need for them to invest in basic infrastructure and services not only related to their activities but which can also serve the communities and help improve the living conditions of rural dwellers. Thus, rural dwellers (especially the youth) can be motivated to stay and serve the rural economy rather than migrate to urban centres where they face the risks of unemployment, crime and other vices. This can be directly beneficial to businesses operating in the rural areas of Africa, through returns in kind for ESG investments.

2.4.7 Social Protection in Africa Generally

According to the International Labour Organization (ILO) (2017), social protection is a set of policies and programmes designed to reduce and prevent poverty and vulnerability throughout the individual's life-cycle. Social protection caters to nine main areas of need: child and family benefits, maternity protection, unemployment support, employment injury benefits, sickness benefits, health protection, old age benefits, disability benefits, and survivors' benefits. These needs are paid for through contributory schemes (social insurance) and/or non-contributory schemes (tax-financed social benefits).[44] In Sub-Saharan Africa, coverage by statutory social security schemes is extremely limited and confined to workers in the formal economy and their families. Sub-Saharan African countries have the lowest coverage of social protection and health care. The ILO put the continent's social protection coverage at 17%, compared to 47% for the global average (ILO 2017; WSPR 2020/22). Sometimes even large private companies come under criticisms for not providing any or adequate social protection for their employees

[43] World Bank (ibid.); and World Bank, *Global Economic Prospects, June 2020*. Washington, DC: World Bank, 2020.

[44] ILO, *World Social Protection Report 2017–2019*. Geneva: ILO, 2017, p. 194; also quoted in Suwun Rita Beri, "Extending Social Security Coverage to the Informal Workers in Cameroon," *Socialprotection.org*, July 2, 2018.

and retirees. ILO data shows that for old age, disability, and survivor pensions, only a little more than 10% of the economically active populations in Ghana and Zambia, and just over 5% in Senegal, for examples, are contributors to pension schemes. These figures mainly reflect the fact that only a small number of workers function in the formal economy, whilst workers in the informal economy are not covered at all. Weak and inefficient delivery of services restrict the coverage of existing social schemes. Furthermore, governance and administrative problems in some existing social security schemes undermine trust and public support for social security. High levels of unemployment and underemployment, as well as the inadequacy of current labour and social protection standards, hamper the effective administration of social protection in many countries.

Social protection enables human development and promotes political stability and peace. Developing effective social protection in fragile African states, in particular, can enhance sustainable development as it protects people throughout their lives against poverty and risks to livelihoods and brings them nearer to their full potential. The utilization of their full potential would make them "assets" rather than liabilities for employers and potential employers in the public and private sectors. Thus, businesses become stakeholders in their well-being, a very strong case for ESG programmes.

As life expectancy at birth increases across the continent, the proportion and number of older persons in the population are also increasing, although overshadowed by the rapid growth of the youthful population. On reaching the age of 40, Africans can now expect to live another 30 years, on average, in all African countries except Botswana, Lesotho, and Swaziland, where the remaining life expectancy at age 40 is less than 20 years because of the prevalence of HIV/AIDS.[45] Older persons constitute a particularly vulnerable population group, as traditional support mechanisms become weakened across Africa because of modernization. Only a small proportion of older persons have access to old age pensions: only about 3% of those aged 65 and older in Burkina Faso, 9% in Ghana, and 10% in Senegal, for example. On average throughout Sub-Saharan Africa, only one out of every five old persons receive old age pensions.[46]

[45] ILO, "Social protection in Africa," https://www.ilo.org/africa/areas-of-work/social-protection/lang--en/index.htm, last visited March 29, 2022.

[46] ILO, *Social Protection for Older Persons: Key Policy Trends and Statistics*. Geneva, 2014.

Formal levels of social protection are naturally lowest in the rural areas throughout Africa. This includes both security services (police and military) and social security. Although the isolation of rural communities from the urban areas and the relatively strong communal ties in the former insulate them from the high crime rates in many African cities, rural areas suffer the most in times of conflicts and natural disasters, precisely because of such isolation and the lack of adequate security services during national emergencies. Social security coverage is also generally non-existent, because of high levels of unemployment and underemployment. The ILO (2018) has pointed out that with 85.8% of employment in Sub-Saharan Africa found in the informal sector, rural communities are made particularly vulnerable to economic and political shocks as well as other crises. This is because of the concentration in small informal enterprises.[47]

Political and legislative commitment, effective leadership and inclusive dialogue are key contributors to the success of social protection and welfare (WHO 2017). However, the private sector in Africa can contribute through voluntary social projects for communities through ESG and respecting at least the minimum legal requirements for their employees and retirees.

Countries all over the world aim to provide social protection for their citizens and the opportunities for all or some of their residents. Therefore, social protection is a global concern and important in any country. Fragile states are in desperate need of adequate social protection schemes but they lack the capacity to provide it (Devereux 2000; ODI et al. 2005). Globally, six out of the ten most fragile states are in Sub-Saharan Africa. These include the Central African Republic, Chad, Democratic Republic of Congo, South Sudan, Somalia, and Zimbabwe. Out of the 30 most fragile countries 21 are in Africa (The Fund for Peace 2019). These states face a myriad of security threats and weak financial and governance capacities to help improve the living conditions of the most vulnerable and marginalized members of their countries. Adopting social protection schemes as tools for dealing with exclusion and poverty, and strengthening institutional capacities to meet social security needs are key to

[47] International Labour Organization (ILO), "More than 60% of the World's Employed Population are in the Informal Economy," *Newsroom*, April 30, 2018, https://www.ilo.org/global/about-the-ilo/newsroom/news/WCMS_627189/lang--en/index.htm, last visited March 24, 2022.

sustainable development in Africa. The latter is a role mainly for government and public agencies, whilst the former is a role that can also be played by the private sector, especially successful companies, through ESG programmes. In other words, implementing social protection schemes at least at a minimum is a role for both companies and governments, but legislating and the strengthening of institutional capacity for doing so nationally is the responsibility of governments. Companies can execute their roles through formal comprehensive ESG programmes.

2.4.8 Micro, Small, and Medium Enterprises (MSMEs)

As noted already, the overwhelming number of businesses in Africa are MSMEs. These enterprises face multitude challenges in their existence. The large and successful companies can help the small ones through their ESG programmes in a variety of ways through their community relations/sustainability departments or through NGOs who work with local MSMEs. From our practical experience running CSR projects in Africa for a number of mining companies two of the most effective ways of supporting MSMEs are to create subcontracting opportunities and organize short-term training for capacity building for them.

However, as mentioned earlier, the main responsibility for MSME empowerment lies with their governments, international organizations, and NGOs. The momentum is picking up in Africa. A number of Africans countries have created whole Ministries dedicated to MSMEs. The African Union Development Agency (AUDA-NEPAD) launched a very ambitious campaign in September 2019 called "100,000 SMEs for 1 million jobs by 2021", with the goal of matching the annual growth in job demand in Africa through the creation of 440 million jobs in the continent by 2030, in line with Agenda 2063.[48] This initiative has gained the support of Ecobank and many other stakeholders, including the World Bank Group, Mckinsey and Company, UNDP, Microsoft, U.S. Mission to

[48] AUDA-NEPAD African union development agency, *Mobilising $1 for the Realisation of $5 towards the benefit of Member States' Project Implementation. Resource Mobilisation Strategy 2021–2024 to Support Africa's Development*. Johannesburg: African Union Development Agency—NEPAD, n.d.; NEPAD, "Launch of 100,000 SME's for 1 Million Jobs by 2021 Campaign," *Africa Renewal*, September 20, 2019; UNIDO, "Report on the Third Industrial Development Decade for Africa (IDDA III) Report by the Director General," March 18, 2021.

the African Union, U.S. Agency for International Development (USAID), and the U.S. International Development Finance Corporation (DFC).[49]

The Arab Bank for Economic Development in Africa (BADEA) also launched the "Africa MSMEs Global Coalition" at the Arab–Africa Trade Forum held in Cairo, Egypt, in November 2021, advocating for a holistic approach that can produce synergy among stakeholders.[50] The Bank has set MSMEs, the youth, and women entrepreneurship as one of the four pillars of BADEA 2030, its strategic plan. To this end, BADEA brought together a global coalition and created a framework where key stakeholders, utilizing the comparative advantage of each, can engage and address the challenges faced by MSMEs. The targeted members of the coalition (some of whom signed the coalition agreement in Cairo) are clusters of MSMEs, specialized agencies of the United Nations, the African Union and its Regional Economic Communities (RECs), Development Finance Institutions (DFIs), banks and other corporations specialized in loans, guarantees and insurance, academics, think-tanks, foundations, and philanthropists.[51] In July 2022, BADEA started ESG training for all its senior staff, which was conducted by the authors of this book.[52] In August 2022, BADEA signed a line of credit agreement with Trade and Development Bank (TDB) in Mauritius to support African SMEs. In August 2022, the authors and ESG Enterprise (a Software as Service (SaaS) and data analytics company for businesses, NGOs, and governments, based in Houston, Texas) for staff in the Bank's Strategy Division.

The International Financial Corporation (IFC) of the World Bank, in particular, has for many years been addressing the financial challenges of MSMEs. The IFC is at the forefront of such efforts since the mid-1990s.

[49] U.S. Mission to the African Union, "USAID, DFC and AUDA-NEPAD Collaborate on the '100,000 MSMEs' Initiative," June 21, 2021, https://www.usau.usmission.gov/msmes-initiative-062121/, April 4, 2022.

[50] ANA, "SME: BADEA launches Africa MSMEs Global Coalition," November 8, 2022, https://www.africanewsagency.fr/sme-badea-launches-africa-msmes-global-coalition/?lang=en.

[51] Sidi O. Tah, "Promoting Micro, Small and Medium Enterprises (MSMEs) in Africa: A Holistic Approach Is a Must!" https://www.theafricareport.com/a-message-from/badea/arab-bank-for-economic-development-in-africa/promoting-micro-small-and-medium-enterprises/, last visited March 26, 2022.

[52] Mariama Sonko, "ESG: Relevance and Applicability in Africa," Webinar for Senior Staff of the Arab Bank for the Economic Development of Africa (BADEA), July 10, 2021.

Its strategies are based on: (1) financial assistance through intermediaries in Africa; (2) non-financial, indirect, institution-building assistance through co-financing by donors of project-development facilities; and (3) providing support to MSMEs in frontier countries with low income, high risk, less developed banking systems, and low private capital inflows.[53] Available data shows that in 2016 the IFC's total committed long-term MSME-focused portfolio in Sub-Saharan Africa was equal to $1.7 billion, including $250 million of long-term finance to client financial intermediaries for MSME support, but excluding $34 million for 33 advisory projects.[54] The AU, BADEA, and IFC are all public international organizations that are promoting economic development in both the public and private sectors in Africa. All these organizations welcome opportunities to work with private companies in Africa. They may also be channels for different kinds of support to private companies in the pursuit of ESG programmes in Africa.

2.4.9 The Resource Curse

Africa's resource-rich countries are known for their poverty and political instability. The former has been dealt with at great length in this chapter. When it comes to political instability, most of the richest countries in the continent, such as Algeria, Angola, Burkina Faso, Chad, the Congos, Libya, Liberia, Niger, Nigeria, South Africa, the two Sudans, Sierra Leone, and Zimbabwe, have all either historically suffered or continue to suffer. They have been subjected to gross inequalities and or wars associated with their natural resource abundance. Thus, they were all at one time or the other examples of the "resource curse" explained in Chapter 1. In spite of economic progress over the last three decades or so, almost all of them (even the most stable among them) continue to face serious economic and political challenges. Companies benefitting from natural resources in these countries can enhance or contribute to political stability in three areas of priorities, namely education, health, and addressing acute poverty. These priorities are overarching and directly

[53] Independent Evaluation Group, *Financing Micro, Small, and Medium Enterprises An Independent Evaluation of IFC's Experience with Financial Intermediaries in Frontier Countries*. Washington, DC: IFC, 2008.

[54] Ferreira (ibid.).

or indirectly embrace all the diverse areas listed in this chapter for ESG actions.

The multidimensional nature of income poverty is expressed in people's lives and livelihoods on a daily basis through the lack of infrastructure and services (discussed in this chapter) and even the inability to fend for themselves in meeting their basic needs, whilst witnessing the abundance of natural resources in their countries. In countries where there are ongoing armed conflicts associated with or fuelled by natural resources (Chad, Mali, Niger, Nigeria, Mali, etc.), the situation of the poor becomes even more desperate. In such countries, it is obligatory for the governments to pursue development strategies to enhance inclusive growth which can address issues of inequality and the paradox of plenty and lift the poor and vulnerable out of deprivation. Again, we must emphasize that this is the role of governments. However, anti-poverty policies in these countries also require some attention from businesses in order to resolve conflicts, mitigate and adapt to climate change, and deal with poverty and related social issues through shared prosperity; in other words they require ESG!

2.5 Governance

As illustrated in the preceding section, MSMEs are the dominant form of enterprises in Africa, In fact, across the developing world, small and medium-scale enterprises (SMEs) account for about 90% of private companies and more than 50% of total formal employment. In Africa, such enterprises, plus micro ones (which make up for almost all of the MSMEs), provide an estimated 80% of total jobs and their importance has been dealt with in detail under the "social" of ESG.[55] Generally, they cannot be expected to have the slightest idea of what formal comprehensive ESG is, in an environment where even government officials and company employees themselves tend not to have much of an idea about ESG (as discussed in Chapter 1 and shown by the private sector survey

[55] See Daniel F. Runde, Conor M. Savoy, and Janina Staguhn, "Supporting Small and Medium Enterprises in Sub-Saharan Africa through Blended Finance," *CSIS Briefs*, July 7, 2021, https://www.csis.org/analysis/supporting-small-and-medium-enterprises-sub-saharan-africa-through-blended-finance#:~:text=In%20the%20developing%20world%2C%20SMEs,important%20driver%20of%20economic%20growth, last visited March 25, 2022.

results in Figs. 2.5, 2.6, and 2.7). However, their situation can be excusable because of their small sizes and revenues and low levels of formal education (if any) of these entrepreneurs. Naturally, they have priorities order than ESG, such as meeting the basic needs of their families. This means that only the relatively few large companies in Africa may be expected to be interested in ESG programmes. For this reason, the number of potential ESG-eligible companies shrinks further because, generally, locally owned companies do not tend to demonstrate environmental, social, and governance concerns, as these are normally not expected of them the same way as they are of the foreign companies, especially the big ones in extractive industries. The results of our surveys of Gambian companies, in different sectors, prove this point (see Figs. 2.5, 2.6, and 2.7).[56] The fact that many African and developing country governments are not in full support of global climate treaties do not help in positively changing the attitudes of local companies in these countries towards environmentally inspired or environmentally focused ESG. That leaves mainly large foreign companies, especially from the western world where ESG is in vogue and in the extractive sectors because of pressures at home and legal obligations in some host country especially when it comes to the environment.

Under such circumstances, the first things to address are governance issues in locally owned companies that are sizable enough in Africa, so that those that have significant environmental and social impacts, can understand and engage in ESG disclosures even at the basic minimum in areas that are material to them. Making governance a priority for such companies (especially the fledgling ones among them) would enable them to build adequate capacity to be run professionally and to understand ESG concerns and address them seriously. Therefore, for such locally owned companies, good governance is what should be sought first and this can be achieved through capacity building and ESG education. This is a role first and foremost for the governments through line ministries and ministries of the environment, trade, and training or other relevant

[56] Data collected and analysed from the beginning of August to the end of October 2022, through questionnaires distributed and completed on the spot by survey assistants to 207 Gambian companies; 41% of the entire registered membership of the Gambia Chamber of Commerce and Industry (GCCI), which statistically represents about 95% level of confidence with 5% margin of error. The survey team was led by Abdoulie D. Bah, Executive Director, African Youth Entrepreneurship Academy (The Gambia).

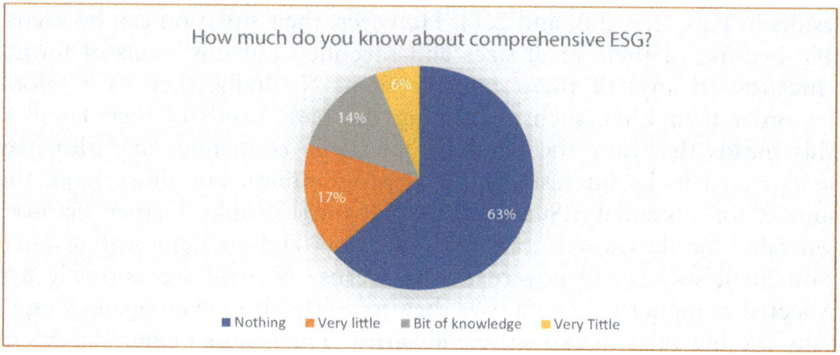

Fig. 2.5 ESG knowledge in the Gambia—Employees of GCCI companies (*Source* Authors)

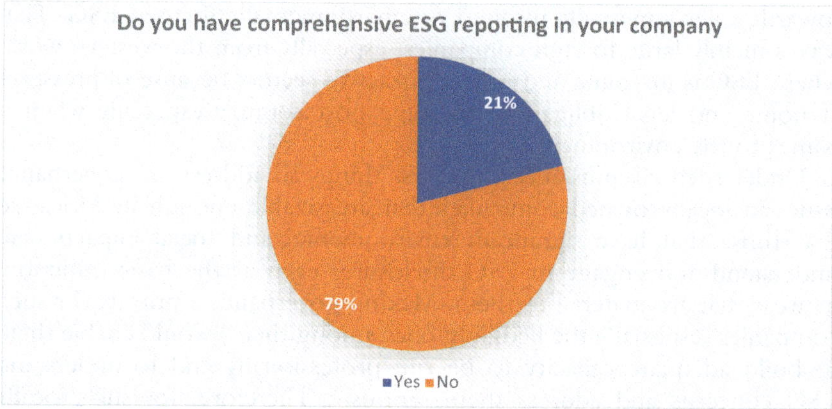

Fig. 2.6 ESG implementation in the Gambia—GCCI companies (*Source* Authors)

ministries. Although we are emphasizing the case or local companies, there is a need for good corporate governance in Africa for companies, whether locally owned or foreign, irrespective of size, as it can help them function more efficiently and enable them to develop the capacity for, and interest in, addressing ESG issues.

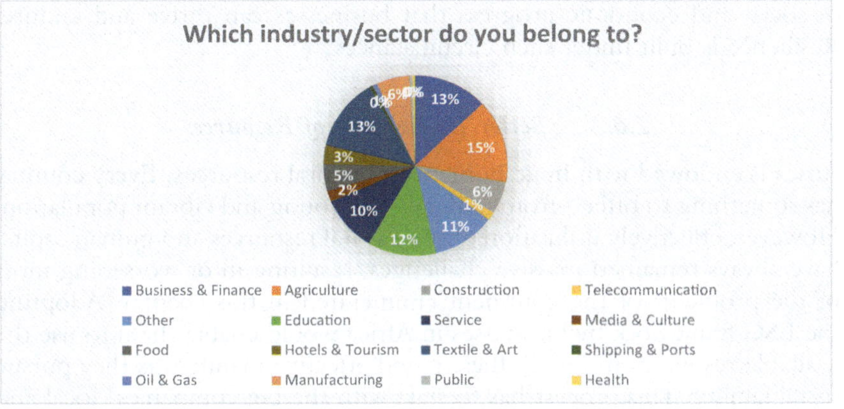

Fig. 2.7 ESG knowledge and implementation in the Gambia—sectors of respondents (*Source* Authors)

2.6 Other Reasons for ESG in Africa

2.6.1 Socio-Economic Progress

In spite of Africa being one of the fast-growing regions in the world, the continent suffers from extreme poverty, as repeatedly illustrated in this chapter. According to the World Bank the number of the extremely poor (worldwide) who depend on $1.90 or less on a daily basis has decreased from 1.9 billion in 1990 to 659 million in 2018. On the other hand, the number of such people has risen in Sub-Saharan Africa, accounting for about 64% of the world's poor in 2018. The Bank has forecasted that by 2030, about 9 out of 10 people in this category will live in Sub-Saharan Africa.[57] As evident from our preceding analysis, ESG can accelerate inclusive socio-economic growth as businesses become more heedful of their workforce, communities, customers, and other stakeholders. It can enhance socio-economic progress and business sustainability, both of which are related because it is in the context

[57] The World Bank, *Poverty and Inequality Platform*, https://pip.worldbank.org/home; Max Roser, and Esteban Ortiz-Ospina, "Global Extreme Poverty" (2013). Published online at OurWorldInData.org. Retrieved from https://ourworldindata.org/extreme-poverty [Online Resource].

of social and economic progress that businesses can thrive and endure. Resilience is built under such circumstances.

2.6.2 *Better Utilization of Resources*

Africa is endowed with huge human and natural resources. Every country has something to offer—from minerals to a young and vibrant population. However, effectively utilization of the natural resources and human capital have always remained massive challenges, resulting in or worsening most of the problems of the continent enumerated in this chapter. Adopting the ESG framework by businesses in Africa would enable them to use the available resources in more efficient and effective manners as they pursue sustainability. This process has to start with the big companies (local and foreign) and the small or sizable ones gradually reined in, on the basis of proposals in the preceding sections.

2.6.3 *Attracting Global Investments*

Africa's investment case is compelling and ESG programmes can make it even more attractive. The Sub-Saharan Africa (SSA) region's gross domestic product (GDP) grew for about 25 consecutive years to 2019, reaching $1.8 trillion that year, a 2.8% increase over 2018.[58] Africa's growth during this period showed that the performance was driven generally by a rapidly growing middle class, better macroeconomic policies, large external capital inflows, and (of particular importance) greater political stability. However, COVID-19 has glaringly exposed the continent's poor health services and vulnerability to external shocks (Chapter 4). Domestic constraints which existed before COVID-19 remain and have become even more acute. These are underpinned by poor national and regional infrastructure, small national markets, and limited access to global markets. The importance of integrating ESG factors into investment analysis and/or running businesses can be a significant means of making investment opportunities more attractive to ESG-conscious

[58] Macrotrends, "Sub-Saharan Africa GDP 1960–2022", based on World Bank data, https://www.macrotrends.net/countries/SSF/sub-saharan-africa-/gdp-gross-domestic-product#:~:text=Sub%2DSaharan%20Africa%20gdp%20for%202019%20was%20%241%2C770.04B%2C%20a,a%206.51%25%20increase%20from%202016, last visited July 12, 2022.

investors and make the private sector a bigger driver of economic progress in Africa.

ESG can be particularly useful for attracting large-scale international financial institutions and partnerships for African business persons. Financial institutions in Africa can also have better access to global capital markets where green investments are increasingly becoming the norm. The availability and efficient and effective usage of financial resources in Africa, with the right attitudes and policies, can create the enabling environment for private sector development. This can make Africa a more desired hub for foreign investments in many industries and boost innovation in the continent. Furthermore, it can lead to more job opportunities, better living standards, reduced poverty, and much more, partly because of and through ESG programmes.[59]

[59] In an analysis, of the energy transition in Africa the International Renewal Energy Agency (IRENA) and the African Development Bank (AfDB) found that an integrated policy framework for the transition could attract a wave of sustainable energy investments to the continent which can in turn boost economic output by 6.4% (*Renewable Energy Market Analysis: Africa and its Regions*. Abu Dhabi: IRENA, 2022).

PART II

The Strategic Management of ESG during and after a Global Pandemic

CHAPTER 3

Applying Strategic Management to ESG: Relevance of a Multipronged Approach Through the African PESTLE Analysis (APA)

Abstract In this chapter, we present the African PESTLE Analysis (APA), an innovative strategic management framework, based on the values of the African woman and deriving from the structure of the traditional PESTLE Analysis. We take the reader through the application and we show how it can be used at the organization and personal levels. We further demonstrate the relevance to ESG.

Keywords African PESTLE Analysis · Performance · SWOT Analysis · Management

3.1 The Modern Company

The global market is extremely competitive! Personal and organizational factors, internal and external, affect the performance of the modern company. The company is faced with many challenges (alternatively "problems"), arising from its internal dynamics as well as, more importantly, the external environment. Therefore, the ability to compete in national and international markets or economies depends on its ability

K. N. M. Sonko and M. Sonko, *Demystifying Environmental, Social and Governance (ESG)*, Palgrave Studies in Impact Finance,
https://doi.org/10.1007/978-3-031-35867-8_3

to deal with both internal and external variables. As stated in Chapter 1, traditionally, the ultimate objective of the modern or any company is profitability. Anything threatening that becomes unwelcomed to investors and managers. However, in today's world the pursuit of profitability alone is problematic for many reasons, as argued in Chapter 1 and briefly explained next.

First, private companies or investors operating in natural resources such as mining, oil, and gas, and even in large-scale agriculture, are constantly facing claims from communities and governments that the resources these companies exploit and/or the land they use belong to the communities where they exist. Therefore, the governments and communities continue to feel that these resources do not belong to the companies, even after the state has legally transferred the titles to the private companies. There continue to be expectations from the governments and the communities that the companies owe them far more than the normal requirements of the law in the form of taxes, salaries and wages, free carried interest, and health, safety and environmental regulations. Therefore, in recent years such companies have come under external pressures and scrutiny which make it difficult for them to focus only on profitability. The larger the company and/or the larger the project, the greater such unwanted attention. We refer to this as the entitlement approach, because the causes of the pressures are due to a sense of entitlement from the state or communities.

Although theirs are generally less than those of natural resource companies, other companies (even those operating in services) now face economic, environmental and social pressures too, indirect if not direct. The direct attacks take the form of criticisms targeting specific companies and even boycotts of their goods and services. Indirect attacks or pressures take the form of general exhortations and psychological statements about corporations which make it difficult, if not impossible, for them too to focus only on profitability. The modern company, no matter in which sector it operates is expected to "give back" to or behave responsibly towards society. Thus, a sense of guilt or responsibility may arise within the company and/or among its investors, especially in the face of local or sometimes countrywide social needs, such as education, health, and economic empowerment. The poorer the community, the greater such a sense of guilt or responsibility may become. We refer to this as the responsibility approach to understanding why companies undertake CSR, II, SRI and ESG. The two approaches form the basis of potential theories to help us to understand the reasons for a company's response to local, national, and international pressures. The second set of reasons, in particular, explain why the modern company willingly or unwillingly, engages in CSR, II, SRI, and ESG, and it complements the analysis we carry out in Chapter 1.

In Chapter 1 we carry out an elaborate narration of international activities within and outside the multilateral system by looking at actions through and by the UN, governments and NGOs. We also explained the global environmental problems which have led to these activities. However, we did not explain theoretically the justifications for such global actions. This has been done in our earlier work three decades ago in which we refer to a "state of global interconnectedness".[1] This state describes the highly interconnected and interdependent nature of the world we live in today. Our world today is very different from that of our forefathers and even that which existed later in the immediate aftermath of World War II. This world functions as a mechanical whole in which the pieces support and depend on each other and a malfunctioning in one part affects the others and the mobility of the entire (global) system, just as robust developments in any part can also benefit the whole (global) system. No longer can we think of maintaining a global economy in which one part thrives whilst the rest die. This is no longer an ideological or moralistic diatribe of politics and diplomacy. The linkages became very clear in the areas of trade and (as discussed in Chapter 1) the environment many decades ago. In trade, studies have demonstrated repercussions effects proving the interconnectedness.[2] In environmental science, research has shown the

[1] This was elaborated in Karamo N. M. Sonko, "Debt in the Eye of a Storm: The African Crisis in a Global Context," *Africa Today*, Vol. 37, No. 4 (4th Qtr., 1990): 15–28.

[2] For very basic explanations of repercussion, backwash or feedback effect the interested reader may refer to Smriti Chand, "Foreign Trade Multiplier: Meaning, Working, Assumption, Explanation, Effects and Criticisms", *YourArticleLibrary* at https://www.yourarticlelibrary.com/macro-economics/theories-macro-economics/foreign-trade-multiplier-meaning-working-assumption-explanation-effect, last visited 26 October 2021; or advanced debates see, for example Paul Krugman, "Growing World Trade: Causes and Consequences," *Brookings Papers on Economic Activity*, Vol. 1 (1995); and the benefits of trade to both industrial and developing countries now, IMF Staff, "Global Trade Liberalization and the Developing Countries," IMF, Washington, DC, November 2001; and for an overview of global trade see, for instance, Esteban Ortiz-Ospina and Diana Beltekian, "Trade and Globalization," Our World in Data, Oxford Martin School, University of Oxford, revised October 2018.

uncontainable worldwide effects of environmental change through global warming, particularly from industrial pollution and deforestation. It was such realities that spurred the global environmental, social and political movements historically, as discussed in Chapter 1. The state of global interconnectedness is stronger today than ever before, as shown by the terribly unhappy evidence of the global consequences of the Coronavirus (COVID-19) pandemic which started in December 2019.

Therefore, in view of the state of global interconnectedness the modern company (unlike its counterpart even about half a century or even less ago) is faced with multiple sources or layers of pressures—governments, communities, NGOs, media, etc., at both the national and international levels.[3] Generally speaking, it does not exist as an island unto itself anymore and to survive it has to be able to deal with the multiple sources of pressure at multiple levels. It can deal with the issues of survival in various ways.

First, it can lobby lawmakers and other influential personalities to shape the laws in its favour, legally, within and/or outside the environment in which it operates. Secondly, it can align itself with the forces of political and/or military power, illegally. Thirdly, it can be a politically neutral player (relative to its predecessor half a century or less ago) within its environment and follow its relevant laws and rules, arising from the national and local levels. In modern democratic societies, characterized by transparency and the freedom of expression, choice, and action, the first and second choices are becoming increasingly complicated for any company. NGOs and (and recently) social media make them even more complicated. Therefore, for sustainable business, dependence on political alliances and complete focus on profitability alone are no longer good ideas for any company. So, the modern company needs to harmonize its political relations and the pursuit of profitability with the expectations and pressures of its environment and, often, even beyond. Therefore, the third option is what is realistic and sustainable. How can this be achieved, particularly in relation to the latter; i.e. the pursuit of profitability?

The best way is to move away from ad hoc measures or guess work to science in its management practices. In management, science means strategy and strategy means science, because management itself is not a science unless it can clearly identify your problems or challenges, their

[3] Pressures from some of these sources might have existed much earlier, but not with the same intensity and in a world as interconnected as it is today.

solutions, actions required to deal with them, assign responsibilities for execution, and, where necessary, specify your targets and deadlines—all systematically and in a disciplined[4] manner.[5] Strategic management frameworks, techniques, or tools are designed to assist companies in this process. Many such tools exist.[6] However, sometimes in their attempts to be scientific they become, unnecessarily, too complicated and impossible for the ordinary manager or supervisor to understand. Furthermore, generally, they do not seem to take into account the effects of personal problems on professional performance and hence personal issues are ignored in work solutions.

Furthermore, none of the frameworks, techniques, or tools integrate a company's challenges using a historic African problem-solving tool with universal appeal and cultural implications, demonstrates the essence of teamwork, offers a comprehensive portrait of the company and its external environment, and bridges the gap between the human/personal and inanimate challenges—simultaneously. This is exactly what the African PESTLE Analysis (APA for short) does, by advancing the conventional Harvard University-inspired mnemonic in a manner that has never been done before and adapting it with flexibility to the needs of any company, institution or individual, anywhere in the world.[7]

[4] Discipline is defined here as orderly or prescribed conduct or pattern of behaviour (Merriam-Webster), https://www.merriam-webster.com/dictionary/discipline, visited February 28, 2022.

[5] According to Kotler (1967), in his review of Aguilar (1967)'s pacesetting work, of all the resources that make up the modern business enterprise—money, materials, machines, men, and information—only information begs for systematic attention from management and business scholars. Despite its increasing cost and value, he observed that most firms have never looked at how to acquire and use external information in a planned manner through "systematic" or strategic management. Philip Kotler, "Review: Scanning the Business Environment. by Francis Joseph Aguilar," *The Journal of Business*, Vol. 40, No. 4 (October, 1967): 537–539.

[6] Other strategic management and marketing frameworks or tools, sometimes referred to as theories, the interested reader may refer to include Ansoff Matrix, Balanced Scorecard, Marketing Mix, Stakeholder Mapping, Porters Five Forces, GE Matrix, Maslow's Hierarchy of Needs, Boston Consulting Group Matrix.

[7] For studies on personal/employee needs and organizational goals the interested reader may refer to A. H. Maslow, "A theory of human motivation", *Psychological Review* (July 1943): 370–396; W. J. Dickson (1973). "Hawthorne experiments," in *The Encyclopedia of Management*, ed. C. Heyel, 2nd ed. (pp. 298–302). New York: Van Nostrand Reinhold; Harpaz, I. "The Importance of Work Goals: An International Perspective," *Journal of International Business Studies*, Vol. 21 (1990): 75–93; F. Herzberg, B. Mausner, and B. B.

3.2 The African PESTLE Analysis and ESG for Africans and non-Africans

In order to deal with the issues of multiple pressures, at the personal and organizational levels, in the modern company, we recommend here a simple framework based on PESTLE Analysis. PESTLE Analysis is normally used by a company or organization in marketing and other activities to analyse and monitor mainly the macro-environmental (external environment) factors that have an impact on it. "External factors are trademarks in the macro environment. The elements show they influence the environment but aren't easily influenced by others".[8]

Harvard Professor, Francis J. Aguilar (1967),[9] is credited for the conventional PESTLE analysis, although the acronym has come a very long way since. In his book, *Scanning the Business Environment*, Aguilar pointed out that "industry is directly and substantially affected by most of the major environmental factors" (e.g., demographic changes, technology, international politics, and so forth) and it is involved in great and continuing changes".[10] He also observed the competition and threats companies pose to each other. Therefore, he came up with the concept of scanning the business environment through macro-environmental factors which he analysed in his book in four main areas at the time—environmental, technological, political, and social. The world outside the firm and dealing with it through planning with strategic management were Aguilar's two main objectives in the book. Later, through different authors and usages, it became known as Political, Economic, Social, and Technological (PEST) or PESTLE (adding Legal and Environmental) analysis; Strategic Trend Evaluation Process (STEP); Social, Technical, Economic, Political, and Ecological (STEPE); Social, Technological, Economic, Ethical, Political, Legal, and Ecological (STEEPLE);

Snyderman, *The Motivation to Work*. New York: John Wiley & Sons, 1959; J. M. Higgins, *The Management Challenge* (2nd ed.). New York: Macmillan, 1994; and K. A. Kovach, "What Motivates Employees? Workers and Supervisors Give Different Answers," *Business Horizons*, Vol. 30 (1987): 58–65; and James R. Lindner, "Understanding Employee Motivation," *Journal of Extension*, Vol. 36, No. 3 (June, 1998).

[8] Kiesha Frue, "Who Invented PEST Analysis And Why It Matters," May 8, 2017. https://pestleanalysis.com/who-invented-pest-analysis/.

[9] F. J. Aguilar, "Scanning the Business Environment," MacMillan Co., New York, 1967.

[10] p. 211.

PESTLE; and other variants.[11] It should be noted that the main interest of most of the authors and users were, and still is, business, especially marketing.

What we refer to as the "African PESTLE Analysis" is an extension of the conventional PESTLE Analysis, which we have done based on our experiences as Africans growing up in a village.[12] The conceptual meanings and literal usages of the pestle by African women in African village environments are utilized within the setting of a modern company. We demonstrate the effectiveness of the African PESTLE in enhancing a company's performance in the pursuit of profitability and attaining the balance required of it in the modern world beyond profitability is shown. In short, the African PESTLE Analysis (Fig. 3.1, now a trademark) used as a tool in a management science or strategic management can help the modern company to identify its challenges/problems in all the six areas, the solutions and how to execute the solutions. As it is shown in the process that is outlined in this chapter, the Analysis fits perfectly with ESG and it can encourage and help any company (non-African or not, involved in African or non-business) to successfully pursue both ESG and profitability. Although we use "company" or "corporation" most of the time in this book, PESTLE analysis can be used in any organization facing challenges and the African PESTLE is particularly suitable for personal challenges too (example divorce[13] and suicide, both of which can be traumatic at home and in the work place).[14]

[11] For a simple but useful narration of the origins of PESTLE Analysis visit https://pestleanalysis.com/who-invented-pest-analysis/.

[12] We are referring to Karamo here.

[13] For works on the causes and effects of divorces see, for example, Piet F. Bracke, Elien Colman, Sara A. A. Symoens, and Lore Van Praag, "Divorce, Divorce Rates, and Professional Care Seeking for Mental Health Problems in Europe: A Cross-Sectional Population-Based Study," *BMC Public Health* (April 29, 2010); Paul R. Amato and Spenser James, "Divorce in Europe and the United States: Commonalities and Differences Across Nations," *Family Science*, Vol. 1, No. 1 (2010); Stephanie Coontz, "The Origins of Modern Divorce," *Family Process*, Vol. 46, No. 1 (March, 2007); and P. C. McKenry and S. J. Price, "Chapter 10. International Divorce," in *Families in Global and Multicultural Perspective*, Sage Thousand Oaks, CA, 2006.

[14] For studies and statistics on suicide internationally, for example Tanya Navaneelan, "Suicide Rates: An Overview," *Statistics Canada* (July, 2012); Danuta wasserman, Qi Cheng, and Guo-Xin Jiang, "Global Suicide Rates Among Young People Aged 15–19," *World Psychiatry*, Vol. 4, No. 2 (June, 2005): 114–120; and A Shah, R. Bhat, S. MacKenzie, and C. Koen, "A Cross-National Study of the Relationship Between Elderly

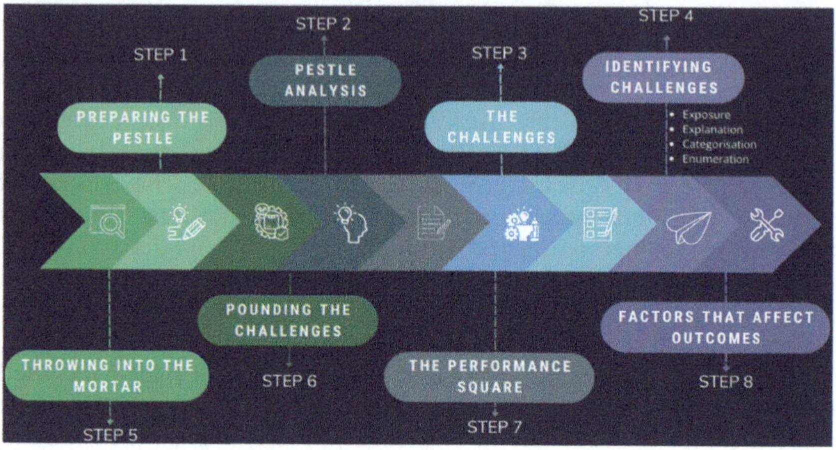

Fig. 3.1 The African PESTLE Analysis™ (*Source* Jula Consultancy FZE, https://apatool.com)

We extend the traditional PESTLE Analysis in the following ways:

1. Utilizing the symbolic and literal meanings of the pestle and applying it to the mnemonic and vice versa—thus pestle for PESTLE and PESTLE for pestle;
2. Explaining the utility of the pestle in feeding/nutrition, problem-solving, travel, transformation, stress management, etc.;
3. Exposing the power of the pestle as a pounder, weapon (self-defence), tool for exercise, etc.;
4. Highlighting the joy from the pestle in traditional festivities and (in some societies) its role during times of mourning (funerals);
5. Categorizing not only the challenges but the pestles themselves by pointing out the importance of choosing the appropriate pestle according to the size of the challenge; and

Suicide Rates and Life Expectancy and Markers of Socioeconomic Status and Health Care," ***International Psychogeriatrics*** (May 16, 2007) at https://www.cambridge.org/core/journals/international-psychogeriatrics/article/abs/crossnational-study-of-the-relationship-between-elderly-suicide-rates-and-life-expectancy-and-markers-of-socioeconomic-status-and-health-care/EC7C16B051B0DC63B321A5BAF048C344, last visited October 26, 2021.

6. Emphasizing the importance of team work as symbolized by the woman, pestle, and mortar.

It uses both traditional strategic management tools (PESTLE and SWOT analyses). We show how the Analysis can be complicated in unique and innovative ways to assist companies and others.

The African PESTLE Analysis is an uncomplicated and enjoyable process of dealing with institutional and personal challenges/problems. In this chapter, the Analysis is employed to ESG, for the first time, in order to enhance profitability and good governance beyond the company level, and achieve social objectives and protection of the environment beyond the company. The relationship between the "the African PESTLE" Analysis and the better-known Strengths, Weaknesses, Opportunities and Threats (SWOT) Analysis is also demonstrated as a further effort to encourage companies to reach the triple bottom line and become better corporate citizens.

Since June 2018, the authors have successfully made numerous presentations of this innovative approach in strategic management to chief executive officers and senior managers at a number of prominent multinational corporations and other institutions in Africa, Asia, and Europe. Outside business, the authors have reached government ministers, ambassadors, and other senior officials serving national and international organizations. Specifically, the APA has been presented at the following institutions, including a triple A-rated multilateral development finance institution, biggest training centre in the Middle East, and one of the biggest international organizations in the world:

1. Arab Bank for the Economic Development of Africa (BADEA) (Khartoum, Sudan)—the Aa2-rated and only Arab multilateral bank financing development across Sub-Saharan Africa.[15]
2. Al Dhabi Investments (Abu Dhabi)—an international macro hedge fund managing about $2 billion in assets.[16]

[15] Karamo N. M. Sonko and Mariama Sonko, "Badea in a New African Economy: Facing the Challenges with PESTLE & Mortar," Presentation to BADEA, May 31, 2016; and Mariama Sonko, "ESG: Relevance and Applicability in Africa," Presentation to BADEA, July 5, 2021.

[16] Karamo NM Sonko & Sonko, Mariama, "Al Dhabi Investments Ltd: Facing the Management Challenges with Arabian PESTLE & Mortar," December 8, 2019.

3. Samil Industrial Co. (Khartoum)—a French-Sudanese agribusiness joint venture which feeds about 2 million malnourished children per annum through the United Nations.[17]
4. Etisalat Academy (Dubai)—the biggest business trainer in the Middle East.[18]
5. Wenso Ltd (Manchester)—a multi-service IT company with blue chip clients.
6. First United Financial and Tax Consultancy (Abu Dhabi)—tax advisors and business consultants.[19]
7. The Secretariat of the Organization of the African, Caribbean, and Pacific States (OACPS) (Brussels).[20]
8. The African Capacity Building Foundation (ACBF) (Harare, Zimbabwe).
9. Ministry of Minerals (Khartoum, Sudan).[21]
10. United Nations Industrial Development Organization Arab International Centre for Entrepreneurship and Investment (UNIDO AICEI) (Bahrain).[22]
11. Meyas Sand Minerals Company Ltd (Khartoum), subsidiary of TSX-listed Orca Gold Inc.[23]

[17] Karamo N. M. Sonko and Mariama Sonko "Samil Industrial Co. Ltd: Facing the Challenges with African PESTLE & Mortar," December 26, 2019.

[18] Karamo N. M. Sonko and Mariama Sonko, "Strategic Management in your Company or Organization: Facing the Challenges with the African PESTLE Analysis," Pilot Session, Etisalat Academy, December 30, 2020.

[19] Karamo N. M. Sonko and Mariama Sonko, "Strategic Management at Corporate & Personal Levels: Facing the Challenges with the African PESTLE & Mortar," Online Presentation to FUFC, Abu Dhabi, Wednesday, July 27, 2020.

[20] Karamo N. M. Sonko and Mariama Sonko, "The African PESTLE Analysis," Online Presentation to the OACPS, Brussels, Belgium, March 26, 2021.

[21] Karamo N. M. Sonko and Mariama Sonko, "Strategic Management: African PESTLE Analysis for African Governments & Regional Organizations," Presentaiton to ACBF, Harare, Zimbabwe, March 19, 2021.

[22] Karamo N. M. Sonko and Mariama Sonko, "Strategic Management in your Company or Organization: Facing the Challenges with the African PESTLE Analysis," Presentation to UNIDO AICEI, Bahrain, May 31, 2021.

[23] Karamo N. M. Sonko, "The African PESTLE for Meyas Sand Minerals Company Ltd," Presentation to Senior Managament Team,, Khartoum, Sudan, February 6, 2020.

12. NextEnergy Capital (London).[24]
13. Geological Research Survey of the Sudan, Ministry of Minerals, and the Sudanese Mineral Resources Company (SMRC), Khartoum.[25]
14. Abu Dhabi Islamic Bank (Sudan subsidiary), Khartoum.[26]
15. Islamic Development Bank, Regional Hub of Dakar, Senegal.[27]

The African PESTLE Analysis is now being developed into a computerized application. This chapter, the first publication of APA, demonstrates its application to ESG. Like this chapter, the chapter is intended more for practitioners than academics. The aim is to help them understand and design their own APA frameworks and apply them to their organizational and personal challenges, including environmental, social, and governance issues.

3.3 The Presentations

The presentations to companies and organizations of the APA take the form of engaging power point presentations that help them identify and address their challenges on their own, rather than dictating them and the solutions. The presentations take place in two stages: (1) A short session to selected relevant staff (typically from the human resources and/or finance departments) during which the process is demonstrated and the challenges are submitted by staff; and (2) Follow-up sessions during which the challenges submitted by (or identified with) the staff are addressed in five categories—political, economic, social, technological,

[24] Karamo N. M. Sonko and Mariama Sonko, "NextEnergy Capital: Facing the Challenges with PESTLE & Mortar," London, Presentation on March 10, 2020.

[25] Karamo N. M. Sonko and Mariama Sonko,"The African PESTLE Analysis," Presentation to GRAS, Ministry of Minerals, Khartoum, Sudan, May 25, 2021; Karamo N. M. Sonko and Mariama Sonko,"Promoting Domestic and Foreign Investment in the Sudanese Mining Sector," Presentation to SMRC, Ministry of Minerals, Khartoum, Sudan, August 11, 2021.

[26] Karamo N. M. Sonko and Mariama Sonko, "Doing Business in Africa: Opportunities, Risks and Strategies," Presentation to the Abu Dhabi Islamic Bank, Khartoum, Sudan, October 11, 2021.

[27] Karamo N. M. Sonko and Mariama Sonko, "Islamic Finance Institutions in an Environment of Fierce Competition and Acute Poverty: Online Strategic Management Presentation to the Islamic Development Bank," December 15, 2021.

and environmental—i.e. the PESTLE. After this, some of the most important threats and weaknesses are plotted into the better-known SWOT format to show the relationship between the PESTLE and SWOT and also identify key strengths and opportunities from the PESTLE Analysis. The length of this follow-up session (s) depends on the number of challenges identified and how much the participants contribute to the discussions. Finally, a strategic management document is prepared with staff and circulated to all relevant managers and supervisors.

During the interactive presentation sessions, it is also shown to staff/participants how the African PESTLE can be used to address their personal problems, which can affect their personal happiness and, thus, work performance. A very short video is shown at the beginning of the presentations, portraying African village women with their pestles and mortars pounding millet.[28] The video demonstrates the following values of the African village woman as means of survival in dealing with numerous challenges in a harsh and restrictive village environment:

Adaptability
Coordination
Confidence
Discipline
Energy
Focus
Patience and Perseverance
Positivity.

Each of the above terms are defined and explained to staff how each can assist them in developing the positive attitudes need to deal with challenges at home and work. The interactive sessions help to make the staff feel part and parcel of the problem-identification and problem-solving processes. The combination of both features (the abilities to address organizational and personal challenges), plus the African dimension, give the African PESTLE Analysis an edge over other strategic management tools.

[28] Any video depicting the African village woman's industry through the pestle and mortar may be usable. Such a video can be viewed, for example, at https://www.youtube.com/watch?v=x9f2PAz94rw.

3.4 The Process

3.4.1 Preparing the PESTLE

The introduction of the African PESTLE starts with explaining what is a pestle. Literally, a pestle is a tool for pounding a substance (food items) in order to crush, refine, and turn them into paste or powder to cook or eat directly. Symbolically, it connotes food, parties, good news, joy, strength, sustenance, direction (a "village positioning system" (VPS) for travellers who are guided by its sound), health, etc. In particular, and of great importance to the modern company, it directly symbolizes *teamwork,* because no matter how strong a pestle is, it can only do its job in "partnership" with a pounder (the woman) and mortar. Without its partners, the powerful pestle is useless in its main "vocation" or task—pounding. With its partners, it becomes robust, overpowering, loud clear, and successful.

In traditional African societies where it is found, the pestle has many uses. Culturally, it is used in all the major festivals (marriages, naming ceremonies) and in funerals in some places to prepare food for guests. Industrially, it is one of agrarian Africa's most common food technologies, which is still irreplaceable by modern technologies because of its convenience (easy to obtain, operate, carry, and store) and market forces which make them affordable to all households. It is easy to manufacture and sell cheaply, compared to grinding machines or other modern technologies which have now made their way across urban Africa but are still largely missing in the rural areas. However, even in the urban areas, pestles or pestle-grounded local foods are still more common in the local markets than those produced by modern tools. In fact, across Africa the most common stable foods, such as rice, maize and other cereals, yams, plantains, green bananas, cassava as well as many vegetables and spices, are pounded or can be pounded before cooking.[29]

The pestle and its teammate (the mortar) provide a natural gymnasium for their teammate—the African woman in the village when they are

[29] For food consumption patterns in West Africa, example, see OECD, Food System Transformations. Implications for People and Policies, OECD, *Maps & Facts*. No. 4, April 2021; and for African stable foods see Afrol News, African Stable Foods, http://www.afrol.com/archive/food_staples.htm, last visited October 26, 2021.

pounding. The movement of pounding with a pestle is similar, in some respects, to the physical activities of weight lifting, but may be more beneficial on certain parts of the body. The movement of pounding can also be helpful in times of frustration, tension, anger, or stress.[30]

Finally, in the past the pestle was convertible into a weapon of self-defence for women in the absence of their men. Even in normal circumstances today in rural areas, it made the women feel powerful and in charge of domestic/home affairs.

In conventional strategic management, PESTLE is a mnemonic or acronym which stands for:

P = POLITICAL.
E = ECONOMIC.
S = SOCIAL.
T = TECHNOLOGICAL.
L = LEGAL.
E = ENVIRONMENTAL.

Actual examples in each category are shown in Table 3.1, drawing from one of the presentations already made to a Sudanese international agribusiness company called Samil Industrial Company Ltd (mentioned above). Before then, we turn to understanding what are challenges next.

3.4.2 *What Are Challenges?*

We refer to challenges or problems first and foremost simply as the "A-Z of life". This means that they exist in all areas, and at all times, of our lives, from birth to death. They are our daily struggles which we do not normally seek or desire but have to deal with by the lack of any other

[30] Research has shown that "all types of exercise can be beneficial for stress management" (Erica Jackson, Stress Relief. The Role of Exercise in Stress Management," *ACSM's Health & Fitness Journal*, Vol. 17, No. 3 (May/June, 2013): 14–19. Medical evidence has shown that pounding can be beneficial. See, for instance, Adetoyeje Y. Oyeyemi, Fati Jajimaji, Adewale L. Oyeyemi, and Abdul-Hameed A. Jabbo, "Cardiovascular Responses to Millet Pounding Activity among Women in a Rural Community in Northeastern Nigeria," *Annals of African Medicine*, Vol. 16, No. 1 (January–March, 2017): 24–29—"This exploratory study suggests pounding activity is a moderately intense physical activity that is includible in exercise and physical activity prescription within the 4–6 METs category" (p. 28).

Table 3.1 List of challenges with actual examples from Samil Industrial Co. Ltd

POLITICAL	*ECONOMIC*	*SOCIAL*	*TECHNOLOGICAL*	*LEGAL*	*ENVIRONMENTAL*
1. Political demonstrations	Inflation	Social/Family events	Low productivity	Restrictive labour law	Waste disposal
2. Changes in government policies	Exchange rate risk	Pregnancies	Inexperience	Frequent changes in laws	Plastic bags
3. Changes in Ministers	High costs of inputs	Ethnic ties	Technical breakdowns		

choice or because of the force of circumstances. Challenges are part of our being and/our environment. Everyone has them, from kings to beggars, babies to grandparents, males and females. In unprofessional, undiplomatic, unofficial, or layman's vocabulary challenges are referred to as, and mean, "problems", "obstacles", or sometimes "dangers".[31] However, not every challenge is a problem or a danger, because some challenges may be voluntarily created to achieve a desired goal that does not pose any problem or danger for us. In professional language, challenges may also refer to threats, weaknesses, opportunities that are difficult to seize, and the tasks required to overcome threats, weaknesses, and difficulties.

3.4.3 *Identification*

After understanding what challenges are, the next step is to identify them. There are two main ways of identification—the process of collecting or acquiring information which defines or describes the challenge and help us perceive it. These ways are:

1. *Exposure*: Directly experiencing or learning about the challenge from first-hand experience. An ache in the head or tooth tells you that you have a headache or toothache. An exploration company looking for gold and finding only subeconomic intercepts for a decade should know it is time they left and looked for another exploration and prospecting licence. They do not need to be told by anyone. Exposure may be incidental, coincidental, spontaneous, or heuristic.
2. *Explanation*: Indirectly learning about a challenge from a third party. For example, an investor who learns through a consultant that there are no roads in an area where he wants to invest in an iron ore project. Another example is when the CEO of a company is told by his/her accountant that the company is falling sort of its profit forecasts and running out of money.

[31] There can be many other terms in layman's language but the main difference between them and the word "challenge" is that a "challenge" reflects a positive attitude and a readiness to face undesirable situations or difficult objectives. Thus, describing something as a challenge rather than a problem or danger implies means they should be expected or considered normal in a professional or professional setting and therefore they can and must be addressed.

What we refer to as identification is similar to what Aguilar refers to as "scanning", which he defines as "exposure to and perception of information". He identifies four modes of the scanning process: undirected view—general exposure to information without specific reason; conditioned viewing or directed exposure—receiving information without actively searching for a clearly defined type of information; formal search—relatively limited or unstructured search for a specific information or reason; and deliberate search—seeking information with a plan, procedure, or methodology, to collect specific information for a specific information or reason. He explained that the type of scanning is determined by the type, value, cost, availability, etc., of the information required, with his focus on strategic information.

Challenges in general can be grouped into types or levels of difficulty or seriousness. Thus, some are "child's play", meaning they are simple to understand and deal with. Some (the second or medium level) are more than child's play, but we can live with, or ignore, them. Some (the third and highest level of difficulty or seriousness) are most urgent or complicated and we cannot live with them, so they require equally serious efforts to deal with. Among these category, some are destructive or even deadly, because of the extremely high levels of understanding abilities and resources required to deal with them. Examples are natural disasters (Tsunamis, hurricanes, etc.) and pandemics (Coronavirus, ebola, etc.).

The ones that must be dealt with can be grouped as follows:

Some that are fun to deal with (example, cigarette smoking!). Some that are *not* fun to deal with (diseases, bankruptcy). Some are simple dilemmas or stalemates (a wrestling or boxing match with no end in sight or winners).

Next, we come to the categorization and enumeration of challenges according to the PESTLE acronym. Categorization provides groups, which enables you to put together the challenges from the same sources or considered to be similar to each other. Thus, the challenges are grouped together according to similarities and separated according to differences. Whatever the case may be, the decision of where to put a challenge depends on how clearly it fits in the definition of any of the six PESTLE categories below or where a decision-maker or team objectively believes it belongs.[32] Examples are shown in tabular form later in this chapter.

POLITICAL

Here we list all the political challenges that should or must be dealt with in order to achieve present and/or future objectives of the company or any other institution.

ECONOMIC

Here we list all the economic challenges that should or must be dealt with in order to achieve present and/or future objectives.

SOCIAL

Here we list all social challenges that should or must be dealt with in order to achieve present and/or future objectives.

TECHNOLOGICAL

Here we list all technological challenges that should or must be dealt with in order to achieve present and/or future objectives.

LEGAL

Here we list all the legal challenges that should or must be dealt with in order to achieve present and/or future objectives.

ENVIRONMENTAL

Here we list all environmental challenges that should or must be dealt with in order to achieve present and/or future objectives.

[32] Subjective decisions are encouraged in this process because of randomness which can hamper the management process. Therefore, if the decision-maker is not sure where to put a challenge the best thing is to ask for the opinion of his/her team.

3.4.4 *Tabulation*

Enumeration is the process of assigning numbers to challenges. Like pictures, numbers give greater precision and clarity than words. Unlike words, they are very clear because they are never ambiguous or ambivalent. A word may have a thousand meanings and synonyms, a number always has one meaning and one synonym. By enabling us to construct lists, enumeration provides clarity to challenges by helping us to group them according to specific criteria. The precision of numbers enables greater organization and targeting.

3.4.5 *Identification*

The identification of challenges is completed only when the challenges are categorized and enumerated. To provide further clarification they are put into one of the simplest but most useful inventions of basic human thinking—tabulation. A table, like a picture, provides clarity through vision by summarizing, organizing, solidifying, and providing precision to numbers, words, and images. When challenges are tabulated in PESTLE Analysis, they get closer to a portrait or photographs in general, by facilitating viewing, understanding, and analysis. Numbers and tables combined constitute a formidable force in strategic thinking. For our purposes, tables are even more useful because the challenges do not have to be listed in any consecutive order.

Examples of challenges faced by Samil Industrial Company Ltd in Khartoum, Sudan, are shown in Table 3.1.

3.4.6 *The Solutions*

Next, we move to the stage of throwing the challenges into the mortar and pounding or crushing them with the pestle/PESTLE. In other words, we address the challenges through actions aimed at finding the right solutions which enable the company to achieve its objectives. Examples are given below, some of which are relevant not only for Samil but for any company or organization working in any part of the world, not only in Africa. It is important to note that often we fail to address our institutional and personal challenges or problems because we do not approach them comprehensively. Generally, people think that once you know your problem and the solution it is over, but no, you are only about half way to the solution. Next, as a good manager you need to identify and execute the action required for each challenge and then the responsibility in carrying out the action. We call this the "quartile approach" (Table 3.2) because these activities (challenge, solution, action, and responsibility) are of equal importance in contributing to a 100% successful outcome in addressing each challenge.

Often, when we are faced with a challenge or a problem, we think that knowing the solution is enough, because it provides the answer we need to deal with the challenge or problem. No, knowing the solution is only part of the journey. First, we need to know what is the challenge, then the solution (s), followed by the action(s) required with clarity to execute the solution (s), and finally who is/are responsible—the responsibility (ies). These four areas of activities are equally important and, therefore, we refer to them as constituting the "quartile approach".

3.5 Performance Square

As shown in tracing its origins, PESTLE Analysis is a flexible tool right from its founding. The African PESTLE takes full advantage of this by introducing additional scope and concepts. These include what we describe next as the "performance square", four more columns as shown in Table 3.3. The performance square may be added to the quartile approach tables shown earlier or may be created separately as follows with examples:

Output refers to what the company can produce to concretely show that the challenge has been addressed. The outputs may be the main objectives clearly portrayed. Measurement is the numeric representation of the output, stated with the precision of numbers as already explained. Targets narrow down or define the goals of the output and, finally, the deadline is the date by which a company plans to complete the execution of the solution (s) or actions required to address the challenge (s). Deadlines give urgency to the targets by setting dates by which they must/should be achieved.

3.6 Extensions

The PESTLE categories can be expanded to include two others, such as demography and security. Although both demographic and security issues can be integrated into the framework as it is, a company may wish

Table 3.2 The quartile approach

Challenges	*Solutions*	*Actions required*	*Responsibilities*
Category (Political)			
1. Political demonstrations	Make up for lost time in production	Compensate for production losses by shifting to non-demonstration hours	Production Manager
2. Changes in government policies	Minimize or avoid the effects of such changes	Implement pre-emptive or cushioning measures	Planning dept. or management team
3. Changes in Ministers	Minimize or avoid the effects of change	Implement anticipatory measures Don't get too close to your Minister	CEO/General Manager
Category (Economic)			
1. Inflation	Avoid/mitigate	Foreign currency measures, supplier contracts	Finance dept.
2. Exchange rate risks	Avoid exchange rate losses	Keep as much revenue in foreign currency as legally possible	Finance dept.
3. High costs of imports	Find reasonable prices	Have different sources of supply	Purchasing Manager/Finance dept.
Category (Social)			
1. Social/Family events	Minimize their impact	Prioritize permissible social events	HR dept.
2. Pregnancies	Women of different ages and marital statuses	Early information for expecting mothers Planning maternity leaves	HR dept.
3. Ethnic ties	Fair and Objective HR decisions	Transparent and competence-based recruitment Diversity in staffing	HR dept.
Category (Technological)			
1. Low productivity	Adapt people to your technology Adapt your technology to people	Appropriate training Appropriate technology	HR and the respective depts.
2. Inexperience	Competent and experienced staff	Recruitment and training	HR dept.

(continued)

Table 3.2 (continued)

Challenges	*Solutions*	*Actions required*	*Responsibilities*
3. Technical breakdowns	Always keep your machines in good order	Import good quality equipment Good maintenance schedule	Technical and procurement depts..
Category (Legal)			
1. Restrictive labour law	Adapt where you must Good employee relations	Good legal advice Good communication with employees and labour ministry	HR and dept., company lawyer
2. Frequent changes in laws	Always expect changes and have flexible outlook	Stay in touch with relevant government offices and be well informed	HR dept., company lawyer
Category (Environmental)			
1.Waste management	Contain your waste	Work with other companies and the Ministry of Trade and Industry to build a landfill Minimize waste Recycle waste	Admin Manager Environment dept. Ministry of Trade and Industry
2. Plastic bags	Recycle them Stop using them	Starting using biodegradable bags Offer/sell waste to plastics factories	Environment dept.

Table 3.3 The performance square

Outputs	*Measurements*	*Targets*	*Deadlines*
1. Proper waste management	Tonnes of waste into landfill/annum	Get other companies and the Ministry of Trade and Industry to contribute \$2 m to build landfill Minimize waste by 2 tonnes/annum Recycle half a tonne of paper and other materials/annum	31 December 2022
2. Increased revenue	Millions of dollars/annum	Expand production capacity by 40% to feed an additional 600,000 malnourished kids	31 December 2020

to address them separately by adding a column for each for better focus, disaggregation, and better attention. If the PESTLE is used as it is, demographic factors may be put in the social category, whilst security issues may be put under the political category. With regard to demography, for example a clothing company may be interested in knowing the demographic composition and changes in its market in order to decide whether it should focus on kids, adults, womens, or mens' clothing. In this case, if demography becomes a distinct category, the PESTLE can become PESTLED or PESTLE-D (with D = demography). Similarly with regard to security, for example a private security provider for a mining company may wish to have greater detail on security issues than anything else in its tasks by assessing the internal environment in the mine site (strikes, theft, fist fighting, speeding, etc.) and the external environment (armed robberies, terrorism, war) in order to know where its focus must be and what type of employees to recruit for the protection of a mine. In this case, if the company decides to make security a separate category, the PESTLE can become PESTLES or PESTLE-S (S = Security).

3.7 Strategy Paper

After going through the challenges of a company with its management and/or other representatives and completing the process of addressing them through the innovative framework of the African PESTLE, we end up with an overall output for the company. This final output takes the form of a Strategy Paper for implementation, which is distributed to all relevant managers and supervisors. The paper is prepared by the company, with our advice if needed, and can be as detailed as the company wishes or has the ability to do.

3.8 Similarities/Differences Between PESTLE and SWOT

SWOT Analysis is well known and used in companies all over the world. However, PESTLE Analysis is not. For the benefit of the reader, the main similarity and difference should be pointed out. The results of PESTLE Analysis or the challenges that are identified through it can be used to identify threats and weaknesses, specifically, which can then be used in a SWOT Analysis to fill in the "weaknesses" and "threats" sections. In addition, as the name shows, SWOT adds the "opportunities" and "strengths". Hence the SWOT acronym or mnemonic stands for "S = strengths, W = weaknesses, O = opportunities, and T = threats. In terms of differences, PESTLE Analysis, even in its traditional usage, provides greater scope for sectoral disaggregation to cater to a much wider range of factors, actions, etc., in a more manageable format. Unlike PESTLE Analysis, SWOT Analysis has been primarily used to look at internal factors, although the opportunities and risks can also be external. In Fig. 3.2 we present the relationship between the two by plotting some of the challenges from the Samil PESTLE Analysis into the SWOT framework. In indirect manners, the identification of the challenges of an institution may offer its strengths. In indirect manners too, the solutions to these challenges may also offer opportunities to the institution. In short, there are direct and indirect relationships between PESTLE and SWOT Analysis.

3.9 Relationship with and Applicability of APA to ESG

Now, let us turn to ESG again, more specifically. As it can be clearly seen from the framework, the APA covers the "E" and "S" of ESG. The "G" can be covered under "Political" because there is a relationship between politics and governance, whether governments and corporations. Nevertheless, where the "G" is focused only or mainly on internal policies or procedures of corporate governance, this category can be covered by "L" for "Legal" in PESTLE Analysis. As demonstrated in this chapter, the conventional PESTLE Analysis by itself can help companies to develop

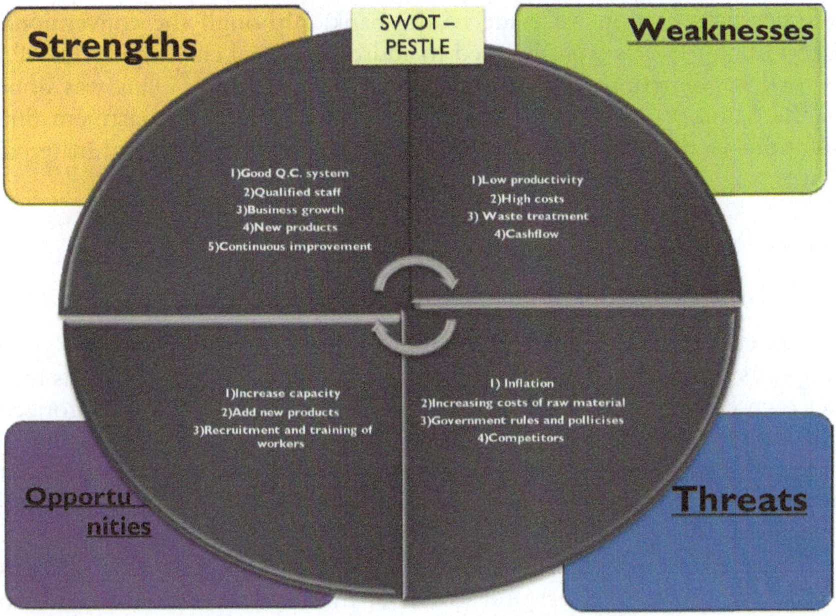

Fig. 3.2 The SWOT–PESTLE relationship (*Source* Authors)

their ESG disclosures or reporting in a comprehensive framework. The APA can expand this framework through further clarity in identifying and organizing the factors, and addressing the challenges that accompany them or of which they represent. Additionality, APA offers Performance Square, which can be used for implementing and monitoring Paper, including the ESG issues.

An example with direct relevance to ESG was shown during the Samil presentation, during which the company realized that it had a bigger environmental issue than previously thought. Therefore, working with the management team during the presentation, it was recommended that neighbouring factories should be approached to contribute money and approach the Ministry of Trade and Industry to build a landfill in the industrial area. During the BADEA presentation (also cited above) the bank realized that it needed a formal strategy for its operations in Africa, which it later developed, including aspects of the environment, social, and governance. It has now initiated efforts to set up an ESG Unit, following another presentation we made to the bank. Although the conventional PESTLE Analysis is typically used to address external challenges, the APA is used to sometimes include internal challenges too.[33] This was done in the Samil presentation to help the human resources department find solutions to the challenges of large numbers of absenteeism and maternal leaves.

3.10 Applicability of the APA to Personal Issues for all

As defined in this chapter, challenges or problems may be referred to simply as the A-Z of life. Like corporate entities, every individual has his/her own challenges. Like corporate entities too, these can be categorized into political, economic, social, legal, technological, and environmental; if not all, some or most of these categories would be applicable to almost anyone. Therefore, the APA is useful for individuals too in organizing their lives.

The inability to organize themselves and deal with personal challenges can be the cause of misery and worse for most individuals. This can lead to self destruction, the worst of which is suicide (Van Praag 2010; Amato and Spenser James 2010; Coontz 2007; and McKenry and Price 2006). Suicide is the third leading cause of deaths among the youth worldwide. According to the World Health Organization (WHO), about a million people commit suicide and 20 times more attempt it each year; on average worldwide, this equals to a global mortality rate of 16 per 100,000, one death every 40 seconds and one attempt every 3 seconds.[34]

High- and low-income countries account for most of all the suicides worldwide, indicating that suicide has no economic boundaries. While

[33] In fact, Aguilar made it clear from the beginning of his book that he is concerned with the world outside the firm and not the internal environment.

[34] World Health Organization. Suicide rates per 100,000 by country, year and sex, last accessed November 5, 2021: http://www.who.int/mental_health/prevention/suicide_rates/en/index.html; Rajiv Radhakrishnan and Chittaranjan Andrade, "Suicide: An Indian Perspective," *Indian Journal of Psychiatry*, Vol. 54, No. 4 (October–December, 2012): 304–319.

the overwhelming number of suicides tend to occur in low- and-middle-income countries (77%), high-income countries have the highest age-standardized suicide rate (10.9 per 100,000), followed by the low-income (9.9 per 100,000).[35]

Sixty per cent of all suicides are committed in Asia: China, India, and Japan account for about 40% of all suicides, according to WHO.[36] In 2019, 116,324, 173,347, and 19,466 deaths occurred in China, India, and Japan, respectively (Figs. 3.3 and 3.4).

WHO further stated that suicide is the 17th leading cause of death in 2019, accounting for 1.3% of all deaths globally.[37] 58% of all suicides

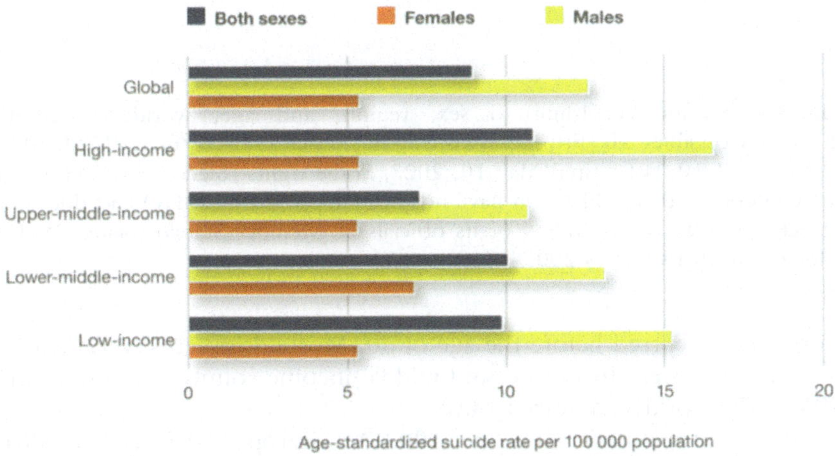

Fig. 3.3 Suicide according to income group, worldwide 2020 (Accessible at https://www.who.int/publications/i/item/9789240026643, pp. 9–10, last visited May 10, 2023. Some rights reserved by WHO and the contents of this publication have not been approved by WHO and does not reflect the views of the WHO or its officials or Member States) *World Bank income groups, 2020 (*Source* WHO Global Health Estimates 2000–2019)

[35] Tosin Philip Oyetunji, M. Yasir Arafat, Stephen Oluwaseyi Famori, Timilehin Blessing Akinboyewa, Michael Afolami, Moyo Faith Ajayi, and Sujita Kumar Kar, "Suicide in Nigeria: Observations from the Content Analysis of Newspapers," *General Psychology* (January, 2021).

[36] WHO, "Suicide data," 16 June 2021.

[37] WHO (ibid.).

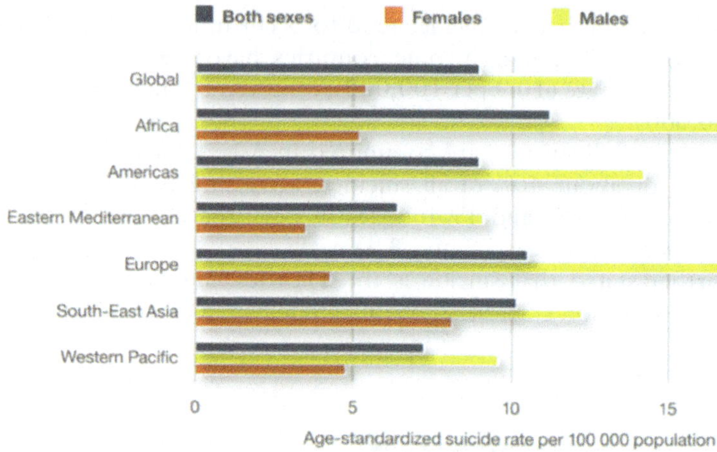

Fig. 3.4 Suicide according to sex, region, and age, worldwide 2000–2019 (Accessible at https://www.who.int/publications/i/item/9789240026643, pp. 9–10, last visited May 10, 2023. Some rights reserved by WHO and the contents of this publication have not been approved by WHO and does not reflect the views of the WHO or its officials or Member States) (*Source* WHO Global Health Estimates 2000–2019)

worldwide occurred before the age of 50. Most adolescents who died by suicide (88%) were from low- and middle-income countries where nearly 90% of the world's adolescents live.

Suicide rates in Africa (11.2 per 100,000), Europe (10.5 per 100,000), and South-East Asia (10.2 per 100,000) were higher than the global average (9.0 per 100,000) in 2019. The lowest suicide rate was in the Eastern Mediterranean (6.4 per 100,000). South-East Asia had a much higher female age-standardized suicide rate (8.1 per 100,000) compared to the global female average (5.4 per 100,000). For males, Africa (18.0 per 100,000), the Americas (14.2 per 100,000), and Europe (17.1 per 100,000) had rates higher than the global male average (12.6 per 100,000).[38]

[38] WHO, *Suicide worldwide in 2019. Global Health Estimates*. WHO, Geneva, 2021.

Studies have found the following among the causes of suicide in the United States, where suicide is the 10th leading cause of deaths (Boston Children Hospital 2021):

- Losses
- Lack of social support or social isolation
- Poor coping skills
- Family disruptions (divorce or problems with the law).[39]

In India, where the suicide rate is about the same as in the United States but increasing, the top 10 causes of suicide have been found to be: family problems (23.7%), illness (21%), unemployment (1.9%), love affairs (2.9%), drug abuse/addiction (2.3%), failure in examinations (1.6%), bankruptcy or sudden change in economic status (2.5%), poverty (2.3%), and dowry disputes (2.3%).[40] In Nigeria, where suicide rates are argued to be grossly under reported as in many other developing countries, the rate of 17.3 per 100,000 is still higher than the global average.[41]

As Oyetunji et al. (ibid.), WHO (2021) and others observed suicide is preventable.[42] Such prevention can be done by the victims concerned or their support networks (for example families) by identifying and addressing their challenges before they get out of hand. That is through a personalized or customized PESTLE Analysis. This does not have to be as formal and elaborate as in this chapter. Instead, it can be a simple informal approach to personal problem-solving based on the principles of the PESTLE, especially the characteristics of the village woman which help her build her survival mechanism. Such a simplified approach can be carried out by the subjects/victims themselves (before things get out of hand) and/or the support network dealing with their problems.

The WHO made the strongest case for the need of an analytical multisectoral tool such as the APA when the organization observed and recommended that: "Suicide is a serious public health problem; however, suicides are preventable with timely, evidence-based and often low-cost

[39] Boston Children's Hospital, Suicide and teens," https://www.childrenshospital.org/conditions-and-treatments/conditions/s/suicide-and-teens, last visited November 5, 2021.

[40] Radhakrishnan and Andrade (ibid.).

[41] Oyetunji et al. (ibid.).

[42] WHO, "Suicide", 17 January 2021.

interventions. For national responses to be effective, *a comprehensive multisectoral suicide prevention strategy* (emphasis mine) is needed".

Divorce is another example. Many studies have been carried out about the causes of divorce, many of which imply the utility of a tool such as the APA (Jiang 2005; Shah, Bhat, MacKenzie, and Koen 2007; S.B. Scott et al. 2013; A.J. Hawkins 2005; X.P. Montenegro 2004; P.R. Amato and D. Previti 2003; C.A. Johnson, et al. 2001; L. Gigy and J.B. Kelly 2001; Thurnher, et al. 1983).[43] Gjelten (2021) summarized these causes as follows[44]:

1. Lack of commitment
2. Incompatibility and growing apart
3. Communication problems
4. Extramarital affairs
5. Substance abuse
6. Domestic abuse
7. Conflicts over family responsibilities.

He noted that 50% of divorces are caused by poor communication and concluded that "while some issues are more harmful than others (like domestic abuse and serious substance abuse disorders), most (causes of divorce) don't have to lead to divorce—as long as both spouses are willing to work together to save the marriage". Therefore, the APA is suitable in

[43] The research results were published in: S. B. Scott et al., "Reasons for Divorce and Recollections of Premarital Intervention: Implications for Improving Relationship Education," *Couple Family Psychology*, Vol. 2, No. 2 (2013); A. J. Hawkins et al., "Reasons for Divorce and Openness to Marital Reconciliation," *Journal of Divorce & Remarriage*, Vol. 53, No. 2 (2012); "With This Ring…", A National Survey on Marriage in America, National Fatherhood Initiative, 2005; X. P. Montenegro, "The Divorce Experience: A Study of Divorce at Midlife and Beyond," AARP, 2004; P. R. Amato and D. Previti, "People's Reasons for Divorcing: Gender, Social Class, the Life Course, and Adjustment," *Journal of Family Issues*, Vol. 24, No. 5 (2003); L. Gigy and J. B. Kelly, "Reasons for Divorce: Perspective of Divorcing Men and Woman", OSU Bureau for Social Research, 2001; C. A. Johnson et al., "Marriage in Oklahoma: 2001 Baseline Statewide Survey on Marriage and Divorce," *Journal of Divorce and Remarriage*, Vol. 18, No. 1–2 (1993); M. Thurnher et al., "Sociodemographic Perspectives on Reasons for Divorce," *Journal of Divorce*, Vol. 6, No. 4 (1983).

[44] E. A. Gelten, "What Causes Divorce? 8 Common Reasons Marriages End", https://www.divorcenet.com/resources/common-reasons-marriages-end.html, last accessed November 5, 2021.

managing marital problems as it enables team work and better communication in addressing challenges. As in the case of suicide, the APA approach in such personal cases do not have to be as formal and elaborate as in the case of institutions and can be carried out by the subjects/victims by themselves (before things get out of hand) and/or the support network dealing with their problems.

It should also be noted that not every category of APA may be relevant at the personal level in all or most cases of both suicidal pressures and marital problems. Whilst the economic, social, and legal categories may very often be necessary, the political, technological, and environmental may not be relevant at the personal level or often. However, the technological may be necessary as modern tools of communication, such as the cell phone and other means of communication can be used to address communication issues in both suicidal behaviour/pressures and marital problems. However, where institutions, such as social welfare departments, hospitals, and law firms, are involved in addressing such problems it may be necessary to use the complete APA in order to address them on a broader or (as WHO (ibid. recommended) multisectoral level.

CHAPTER 4

The COVID-19 Pandemic and ESG Reporting Priorities: Far Beyond Business as Usual

Abstract This chapter assesses the impact of Covid-19 on African economies and the implications for ESG reporting. In the process, we discuss the incidence of COVID-19, the measures that have been or are being undertaken to contain and eliminate it in the continent. We also make recommendations for broader measures as well as ESG-specific actions necessary for the purposes of containment and elimination.

Keywords COVID-19 Pandemic · Africa · Global · ESG · CSR

4.1 Introduction

In this chapter, we assess the impact of Covid-19 on African economies and the implications for ESG reporting. In the process, we discuss the incidence of COVID-19, the measures that have been or are being undertaken to contain and eliminate it. We also make actionable recommendations for broad measures as well as ESG-specific actions for companies that are necessary for the purposes of containment and elimination in Africa, in particular, and the world, in general. Our recommendations are made with expectations of non-traditional business attitudes and operations which see more than profit. The complexities of understanding

K. N. M. Sonko and M. Sonko, *Demystifying Environmental, Social and Governance (ESG)*, Palgrave Studies in Impact Finance,
https://doi.org/10.1007/978-3-031-35867-8_4

the quantitative realities of the COVID-19 pandemic are discussed by examining the common measures used to determine numbers in absolute and comparative terms. The importance of international coordination in dealing with not only COVID-19 but other present and future outbreaks of globally infectious diseases is stressed and the relevance of APA in such coordination is illustrated.

4.2 COVID-19 in Africa

First, let us summarize what we know and what we do not know about the COVID-19 pandemic in Africa up to this stage. In late 2019, a novel coronavirus, identified as SARS-CoV-2, was found to cause a pandemic of respiratory illness called COVID-19. Africa has not been spared from the global economic meltdown that followed. As can be seen from Table 4.1, the pandemic has resulted in a global recession reflected in a historic economic slump in output of 3.1% in 2020, followed by a steep rise of 6.0% in 2021. These rates have been forecasted to be followed in 2022 and 2023 by about half the growth rate in 2021. In Sub-Saharan Africa, the combined GDP fell by −1.7% in 2020, then rose by 4.7, and forecasted by 3.6 and 3.7% in 2022 and 2023, respectively. Although IMF data puts the Middle East and North Africa (MENA) together, the negative effects of the pandemic in 2019 were universal in all the regions, including MENA as shown below in the table. A study by the Fund found that the pandemic has created the risks of wiping out some of the progress made in MENA over the past decades in poverty reduction, literacy, and lifespans and to exacerbate inequality in a durable way.[1] North African countries, which comprise mainly non-oil exporters,[2] are bearing the brunt of COVID-19, unlike their MENA oil exporting counterparts, which were relatively better placed prior to the pandemic

[1] Roberto Cardarelli, Mercedes Vera-Martín, and Mr. Subir Lall (eds.), *Promoting Inclusive Growth in the Middle East and North Africa Challenges and Opportunities in a Post-Pandemic World*, IMF, October 2022.

[2] Depending on how many countries are counted or count themselves as North African.

Table 4.1 The global economy before and during the COVID-19 pandemic

Real GDP, Annual Percent Change	*2019*	*2020*	*2021*	*2022*	*2023*
World Output	2.9	−3.1	6.0	3.2	2.7
Advanced Economies	1.7	−4.5	5.2	2.4	1.1
Emerging Markets and Developing Economies	3.7	−2.1	6.6	3.7	3.7
Emerging and Developing Asia	5.5	−0.8	7.2	4.4	4.9
Emerging and Developing Europe	2.1	−2.0	6.8	0.0	0.6
Latin America and the Caribbean	0.1	−7.0	6.8	3.5	1.7
Middle East and Central Asia	1.0	−2.8	4.5	5.0	3.6
Sub-Saharan Africa	3.1	−1.7	4.7	3.6	3.7

Source of data IMF, World Economic Outlook, October 2022

and which benefitted from the oil price windfall following the start and escalation of the Russian-Ukraine war.[3]

According to the latest regional data from the World Health Organization (WHO), Africa has had the lowest cumulative number of cases and deaths, compared to all other regions, since the beginning of the COVID-19 pandemic (see Table 4.1). The number of cases (9,390,554) and deaths (174,993) were lower than even the United Kingdom's alone (24,000,101 and 196,821, respectively).[4]

[3] For the petroleum price movements and impact on MENA during the pandemic see, for instance: David Gaffen, "Analysis: Oil's Journey from Worthless in the Pandemic to $100 a Barrel," Reuters, February 24, 2022, https://www.reuters.com/business/energy/oils-journey-worthless-pandemic-100-barrel-2022-02-24/; Cheima Gharib, Salma Mefteh-Wali, Vanessa Serret, and Sami Ben Jabeur, "Impact of COVID-19 Pandemic on Crude Oil Prices: Evidence from Econophysics Approach," *Resource Policy* (December 2021); 74: 102392; Published online 2021 October 4. https://doi.org/10.1016/j.resourpol.2021.102392; and Bloomberg, "Oil's Spectacular Covid Crash Set the World Up for $100 Crude," February 17, 2022, https://www.bloomberg.com/news/articles/2022-02-16/oil-prices-near-100-as-global-economy-struggles-to-balance-post-covid-crash. For the impacts on exporters, importers and the energy transition, refer to, for instance: Will Horner, "North African Countries Snap Up Russian Oil Products Shunned by West," *The Wall Street Journal*, February 25, 2023; Callee Davis, "Ripple effects of a Russia-Ukraine war," Oxford Economics, March 3, 2022; Paul Rivlin and Joel D. Parker, "The Ukraine War and the Middle East: The Rich Get Richer and the Poor Get Poorer," *Middle East Economy*, Vol. 12, No. 3 (July 2022); and Fernando C. Hernandez, "Russia/Ukraine War Puts the Energy Transition in Danger," *Journal of Petroleum Technology* (July 5, 2022).

[4] https://covid19.who.int/table/, last visited on November 26, 2022.

Most of the casualties (confirmed cases and the reported deceased) were within the age range of 60 years and above.[5] These ranges and those with underlying medical conditions, such as cardiovascular diseases, diabetes, chronic respiratory diseases, or cancers are more vulnerable to develop serious or fatal illnesses, although evidence shows that anyone, of any age, can indeed die of a COVID-19 infection. When we look data at data for 205 countries, accounting for 99.9% of all COVID-19 confirmed cases and reported deaths, we find that 57% of these countries provide sex-disaggregated data for cases and/or reported deaths, and 41% provide data for both cases and reported deaths. The data shows that men account for 51% of cases, 55% of hospitalizations (55%), 63% of ICU admissions, and 57% of deaths in Africa. Unlike during the early period of the pandemic, almost all African countries now provide sex-disaggregated data on confirmed cases and 21 African countries provide such data on reported deaths.[6]

According to the INFORM Severity Index,[7] only 18 African countries are classified as having very high levels of severity of the COVID-19 pandemic. However, data for the calculation of this composite Index was not available for about half of the total number of African countries.[8]

[5] This is a common trend all over the world. In Japan, for example, the cumulative number of deaths by age groups starts to climb in the 60s for both males and females and is largest in the 80–90-year cohort for both sexes. However, the number for those over 90 is lower than the 80–90-year cohort. https://covid19.mhlw.go.jp/en/. From the Japanese Ministry of Health, Labour and Welfare, see https://covid19.mhlw.go.jp/en/, last visited January 16, 2023.

[6] See https://globalhealth5050.org/the-sex-gender-and-covid-19-project/the-data-tracker/?explore=variable and https://globalhealth5050.org/the-sex-gender-and-covid-19-project/the-data-tracker/?explore=variable. Also see https://data.humdata.org/visualization/covid19-humanitarian-operations/?layer=covid-19_cases_and_deaths_(sex_disaggregated)#.

[7] The INFORM Severity Index is a composite index, which puts together 31 core indicators, in three groups: impact, conditions of affected people, and complexity. The indicators are calculated on a scale of 1 to 5. The scores are then aggregated into the three groups or components to determine the overall severity, with scores ranging from very low to high. The three dimensions are weighted according to their contribution to severity through the impact of the crisis, conditions of affected people, and complexity. More information on the Index can be found in https://www.acaps.org/methodology/severity#:~:text=The%20INFORM%20Severity%20Index%20is%20a%20composite%20index%2C%20which%20brings,scale%20of%201%20to%205, last visited January 16, 2023.

[8] https://data.humdata.org/visualization/covid19-humanitarian-operations/?layer=inform_severity_index, last visited January 16, 2023.

4.3 Anti-COVID-19 Measures

COVID-19 caught the world off guard and there was panic across the world largely because the world was put in shock as no one was prepared for anything like it.[9] There was a dramatic rise in the number of cases at the beginning of what first looked like an occasional outbreak of a flu-like new or old sickness, then a global warning from the medical profession that there was an emergent epidemy caused by a new disease called COVID-19. Finally, as reality sank in, the international medical community declared the COVID-19 pandemic. In the panic mode, the responses to the pandemic by authorities around the world, including African governments, have taken, and some of which still remain in some countries, to be in the following forms:

1. *Detection*. Increases in screening and treatment;
2. *Food*. Hoarding food and feeding underprivileged communities;
3. *Aid*. International development finance institutions, such as the International Monetary Fund (IMF) and World Bank, dishing out grants to the underdeveloped countries to help them to undertake emergency measures;
4. *Information*. Public awareness campaigns;
5. *Domestic Movements*. Restrictions on domestic movement and public transport. Stay-at-homes and curfews;
6. *Masks*. Facial coverings and/or wearing masks
7. *Schools*. Adaptation or closure of schools
8. *Businesses*. Adaptation, closure of businesses, and massive layoffs
9. *Gatherings*. Limits on or restrictions on public and private gatherings
10. *International Travel*. International travel restrictions or lockdowns (entry restrictions, quarantining, and testing).[10]

[9] For the economic impacts of COVID-19 on the world and the responses, the reader also refer to the World Bank, "The Economic Impacts of the COVID-19 Crisis," chapter 1, *World Development Report 2022*, last accessed on April 12, 2023, at https://www.worldbank.org/en/publication/wdr2022/brief/chapter-1-introduction-the-economic-impacts-of-the-covid-19-crisis.

[10] https://covid19.who.int/region/amro/country/ca/measures, last visited November 11, 2022.

At the height of the pandemic the above containment measures cost the continent an estimated US$ 13.8 billion in lost GDP each month and 40 million people were thrown into extreme poverty.[11] The continent and most of the underdeveloped countries started to drum for funds right from the beginning of the pandemic. In Africa, the then Chairperson of the African Union, Cyril Ramaphosa of South Africa, plus the former President of the African Development Bank (AfDB), Donald Kaberuka, the former Nigerian Minister of Finance and now Director General of the World Trade Organization, Ngozi Nkonjo-Iweala, the former South African Finance Minister, Trevor Manuel, and the former CEO of Credit Suisse, Tidjane Thiam, were appointed Special Envoys to mobilize international support for the fight against Africa's COVID-19-induced economic woes. This was not the first time for such action in Africa—the African Development Bank (AfDB), Organization of African Unity (OAU), UN Economic Commission for Africa (ECA), and the New Partnership for Africa's Development (NEPAD) are all well known for their frequent calls for financial resources for themselves and Africa. Nevertheless, past experience also strongly shows the failure of aid at national and even regional levels because of the lack of accountability and other reasons that have been widely studied. The best way to make sure that funds mobilized for containment measures against COVID-19 are used exactly for the purposes that they are raised is to ensure there is national and global level accountability for every cent raised. Many would argue that this is too late as scandals have already hit a number of African countries in their use of COVID-related aid.[12]

[11] World Bank, "The World Bank in Africa. COVID-19 (Coronavirus) Response," October 1, 2021, https://www.worldbank.org/en/region/afr/coronavirus, last visited December 10, 2022. The number of those thrown into extreme poverty has been estimated at 53 million in 2020 by Etayibtalnam Koudjom, Sévérin Tamwo, and Koffi D. Kpognon, "Does Poverty Increase COVID-19 in Africa? A Cross-Country Analysis," *Health Economics Review*, Vol. 12, Article No. 51 (2022).

[12] See, for examples, APA News, "Gambian Minister Makes Sensational Covid-19 Corruption Claims," May 17, 2020; BBC News, "Coronavirus in South Africa: Misuse of Covid-19 Funds 'Frightening'," September 2, 2020; and Africanews, "Africa's Covid-19 Corruption That Outweighs Pandemic," May 25, 2021; last visited December 10, 2022.

4.4 Africa's COVID-19 Mystery

According to the above statistics and common belief within and outside Africa, COVID-19 is not an African disease, both in terms of prevalence and origin. This is a huge relief for Africans because, until COVID-19, Africa and Africans were generally seen as the main source of the world's epidemics, such as HIV-AIDS, EBOLA, dengue fever, etc. Various explanations have been expounded for why the coronavirus is an exception. First, the available evidence has clearly shown that China is the source of the coronavirus, and this is universally proven and accepted. Secondly, according to research evidence the virus cannot survive long in hot weather beyond certain temperature, such as exist across the continent.[13] Thirdly, the availability of space, particularly in the rural areas, restricts the congestion and other urban lifestyles, such as close contacts, apartment dwellings, and ghettos, which create conditions for the coronavirus to thrive in the big cities of other parts of the world. Fourthly, the fact that about 60% of Africa's population is less than 25 years of age has reinforced the continent's ability to resist the virus. As revealed already across, most of the cases and reported deaths tended to be of much older people. Finally, many ordinary Africans or laymen tend to popularly or informally suggest that the genetic composition of Africans or their adaptability to difficult circumstances make them less vulnerable to any germ and virus.

It may be true that the COVID-19 pandemic is less severe in Africa, compared to other regions/continents. It is difficult to argue or prove otherwise when both statistical evidence and common belief say so. However, based on the realities of African economies and societies/cultures generally, one can argue credibly that the severity of the COVID-19 pandemic may be underestimated, as pointed out in some studies.[14]

[13] Medical researchers are not in total agreement. For a summary of the research, and summary of the research and evidence refer to: Hilary Guite, "How Does Weather Affect COVID-19?" *Medical News Today*, August 16, 2020; Domagoj Kifer et al., "Effects of Environmental Factors on Severity and Mortality of COVID-19," *Frontiers in Medicine*, January 20, 2021, https://doi.org/10.3389/fmed.2020.607786.

[14] For example Omer et al. (2021) observed that the mortality rate among Sudanese patients with COVID-19 infection was 29%, this was higher than the reported rates of 17.6% and 3.9% in published studies. However, this rate was still lower than the rate reported in other places, citing Brazil specifically. Ghada Omer Hamad Abd El-Raheem, Maysoun Ahmed Awad Yousif, Doaa Salih Ibrahim Mohamed, "Prediction of COVID-19 Mortality Among Hospitalized Patients in Sudan," MedRxiv, https://www.medrxiv.org/content/10.1101/2021.03.09.21253179v1.full-text.

First, no one can be a 100% certain whether the low numbers of cases and reported deaths across Africa are actual reflections of the reality or indications of the poor medical facilities and/or insufficient reporting in the continent. Most people in Africa do not have access to proper modern medical facilities and many shun it even where such facilities exist in preference for traditional medicine, which make it impossible to detect and report all cases and, in particular, deaths. Secondly, conspiracy theories about COVID-19 abound in Africa (although not only restricted to the continent) and many people, even educated/literate ones, including political leaders, simply do not believe that COVID-19 exists. Naturally, such conspiracy theorists, would not believe that the cause of deaths is COVID-19 even if people die in front of them because of the virus.[15] Thirdly, the preceding two reasons can result in relatively small numbers of tests and declarations. Finally, and again because of these two reasons, post-mortems of the deceased are very rare in Africa and often deaths are assumed to be due to natural or other (non-coronavirus related) causes. The overall effects of all these reasons is to underestimate the severity of the pandemic in Africa. Furthermore, Mathieu et al. (2020) pointed out that it is the case fatality rate that is popularly used when the number of cases and the risk of death from COVID-19 are formally discussed, which has a number of shortcomings.[16] This can be simply illustrated:

$$\mathrm{CFR}_t = \frac{\mathrm{NDD}_t}{\mathrm{NDC}_t} \times 100$$

where:

CFR_t = Case Fatality Rate at time t.

NDDt = Number of deaths caused by the disease at time t.

NDCt = Number of cases of the disease diagnosed at time t.

If 2000 had caught the COVID-19 virus in a country but only 1000 people went to the hospital and were diagnosed with the disease and

[15] See evidence from Nigeria and Tanzania in: https://www.tandfonline.com/doi/full/10.1080/19371918.2022.2048936; https://www.ncbi.nlm.nih.gov/pmc/articles/PMC9210122/; https://www.ncbi.nlm.nih.gov/pmc/articles/PMC9188000/; and https://www.npr.org/2021/03/17/978336051/tanzania-president-john-magufuli-a-covid-19-skeptic-has-died.

[16] Eduardo Mathies, "Mortality Risk of COVID-19," Our World in Data, 2020, last visited December 4, 2022, https://ourworldindata.org/mortality-risk-covid; also refer to Worldometer, "Coronavirus (COVID-19) Mortality Rate," https://www.worldometers.info/coronavirus/coronavirus-death-rate/, last visited December 13, 2022.

reported and 100 people died, the CFR would be $100/1000 \times 100 = 10\%$. If, however, only 500 caught the virus in another country and all of those cases went to the hospital and reported and 100 died, the CFR would be double the one in the underreporting country. Therefore, the CFR depends entirely on the reported cases. In situations where cases are not all or mostly reported, it is impossible to catch the true severity of the disease in any country or place.

Therefore, it follows that if the actual numbers are always reported in a country, the United Kingdom as example only, the absolute numbers as well as the indices would show greater severity in the United Kingdom than in the country where the levels of reporting are always low. Similarly, the COVID-19 pandemic in itself, when compared to any other outbreak of a different disease with a higher reporting rate, would appear a less severe virus because its real fatalities are not truly reflected by the statistics.

Another simple measure for discussing the severity of COVID-19 among countries is the crude mortality rate or the crude death rate. This rate is used to measure the share of deaths in the entire population caused by a particular disease. It is calculated by dividing the number of deaths of a disease by the total population. Obviously, if the number of COVID-19 deaths reported are high in a country, the crude death rate would be high too, if not it would be low without reflecting the reality.

In order to deal with data challenges such as those in the case of Africa and of countries generally during the COVID-19 pandemic, the concept of excess mortality is considered by many to be more suitable. Karlinsky and Kobak (2021) came up with more advanced methodology to measure this as the increase in all-cause mortality in relation to the expected mortality and built a repository for 103 countries and territories. Their World Mortality Dataset of weekly, monthly, or quarterly data, is issued to determine excess mortality in a country during a pandemic, COVID-19 purposely. Their model for calculating the excess is mainly based on four steps/equations summarized here.[17] They estimated the

[17] For further and detailed information refer to: Ariel Karlinsky and Dmitry Kobak, "Tracking Excess Mortality Across Countries During the COVID-19 Pandemic with the World Mortality Dataset," published online June 30, 2021, https://doi.org/10.7554/eLife.69336, retrieved April 1, 2023; Charlie Giattino et al., "Excess Mortality During the Coronavirus Pandemic (COVID-19)," https://ourworldindata.org/excess-mortality-covid#excess-mortality-our-data-sources, April 1, 2023; Lone Simonsen, and Cecile Viboud, "A Comprehensive Look at the COVID-19 Pandemic Death Toll,"

expected or baseline mortality for the year 2020 using available historical data from 2015 to 2019:

$$D_{t,Y} = \alpha_t + \beta \cdot Y + \epsilon \tag{4.1}$$

where $D_{t,Y}$ = the number of deaths observed in week, month, or quarter t in year Υ, β = is a linear slope across years, α_t = separate intercepts (fixed effects) for each week, month, or quarter, and $\epsilon \sim \mathrm{N}(0, \sigma^2)$ = Gaussian noise.

They took the model prediction for the year 2020 as the baseline for excess mortality computations:

$$\widehat{B}\mathrm{t} = \hat{\alpha}\mathrm{t} + \widehat{\beta} \cdot 2020 \tag{4.2}$$

They obtained the final estimate of the excess mortality:

$$\Delta = \sum_{t \geq t_1} (D_{t,2020} - \widehat{B}\mathrm{t}) + \sum_t (D_{t,2021} - \widehat{B}\mathrm{t}), \tag{4.3}$$

where t_1 = the beginning of summation in 2020.

They calculated the "predictive" variance:

$$\mathrm{Var}[\Delta] = \mathrm{Var}\left[\sum_t \mathrm{wt}\hat{B}\mathrm{t}\right] + \sum_t \mathrm{wt}\hat{\sigma}^2 = \mathrm{w^T S w} + \hat{\sigma}^2$$
$$\|\mathrm{w}\|1, \tag{4.4}$$

where the first term = the uncertainty of $\widehat{B}$t and the second = the additive Gaussian noise that 2020–2021 observations would have had on top of B_t if the pandemic event had not taken place.

In spite of the more sophisticated attempts of Karlinsky and Dmitry Kobak, which is now being used or referred to in many studies, the African questions remain. In fact, they reported that no historic data was available for South Africa, expectedly the most data-advanced African country. Even outside Africa, however, they found that the excess mortality was more than 50% of the expected annual mortality or more than 400 excess deaths per 100,000 population.

published online August 12, 2021. https://doi.org/10.7554/eLife.71974, https://www.ncbi.nlm.nih.gov/pmc/articles/PMC8360646/, April 1, 2023.

In short, Africa may be the least affected continent, but the actual figures will never be known. In any case, the pandemic is a reality that has real effects, with at least 174,993 dead in the continent. Furthermore, it might have significantly subsided or even ended in the theory now, but, practically, the threat remains, not only of the coronavirus but other viruses in a world that has become increasingly prone to epidemics and pandemics.

4.5 The Pandemic Might Be Heading to an End, but the Threat Is Not

The COVID-19 pandemic might be heading for an end, or have already ended according to many, but the threat of COVID-19 has not as cases and reported deaths continue around the world. In the 28 days ending on 28 November 2022, there were 11,101,464 cases and 39,991 deaths worldwide.[18] Newly reported cases in Africa in the last seven days during the same month were 4407.[19] As far as this threat remains, even outside Africa, the continent has to build resilience because of the nature of the world we live in today. As we have promulgated in earlier work, our world is no longer the world of our forefathers, not even the world that existed in the peace and reunification years following the Second World War. Instead, we live in a much more interdependent and interconnected world which functions as a mechanical rather than an organic whole. A malfunctioning in each component of today's world affects and sends viral currents throughout the global system, as vibrations in any units stimulate activity in the rusty parts no matter how far away they may be physically.[20] With the COVID-19 pandemic, this is neither a theory nor a mere plea to emotions or even morals and ethics. At no time in recorded human history has the singularity of our interdependence and ability to survive as a world (all or none) become truer. We describe this reality as the "state

[18] https://coronavirus.jhu.edu/map.html.

[19] https://covid19.who.int/table, last visited November 29, 2022.

[20] Karamo N. M. Sonko, "Debt in the Eye of a Storm: The African Crisis in a Global Context," *Africa Today*, Vol. 9, No. 10 (1990). For the risks of the Corona virus in such an environment, see for instance: A. Wilder-Smith, "COVID-19 in Comparison with Other Emerging Viral Diseases: Risk of Geographic Spread via Travel," *Tropical Diseases, Travel Medicine and Vaccines*, Vol. 7, No. 3 (2021), https://doi.org/10.1186/s40794-020-00129-9.

of global interconnectedness". The Coronavirus has made this glaringly true as an indiscriminate overpowering common enemy whose target is anyone with two legs and two hands. Therefore, as far as any part of the world is threatened by the virus, Africa is threatened too and measures to prevent and cure the disease and deal with its inevitable consequences have to be taken very seriously by Africans.

4.6 Global Measures for Dealing with COVID-19

It follows from the above section that to contain COVID-19's direct and indirect consequences, we must behave like and prepare for the long haul, act with speed, coordination, and for the common good on a global scale. Actions at the national and personal levels are great, but cannot contain the disease and create a better world in the long run, unless the global aspects become equally or more important.

Over the last few decades there has been a lot of talk about "the global village" but, in reality, every government (except those who do not care much about their people) acted, and continue to act, in its own national interest, even when it fights for others. No effort is made to hide this fact. COVID-19 has seriously challenged the predominance of such self-interest in international relations. We can defeat it and such global epidemics and pandemics only by holding hands as one world, one global village and community indeed!

Although the need for global solutions is already being felt due to the coronavirus, we are very far from implementing these solutions in manners that can: (1) build preparedness for the long term to be able to deal with such pandemics and epidemics in better ways and (2) create a more just world for all. In effect, we are implying coordinated health measures (an ESG issue) at the global level which goes beyond the company levels.

Concrete actions through strategic management, as outlined in Chapter 2, are urgently required. Our strategic management framework (that is, the APA) is as relevant to Africa as it is to the rest of the world. As demonstrated in Chapter 2, it can help multinational and national companies, to deal with macro-environmental factors in order to survive and become more competitive. It can also help governments through identification and prioritization of the world's challenges that they are responsible for dealing with as well as the solutions and actions required in addressing these challenges. It can go further to help them define outputs, targets, and deadlines.

Sometimes, if not always, governments need to think like private companies and function with the efficiency of private companies, but for the benefit of the global, instead of the private, good. Moreover, the African PESTLE Analysis can help to generate focus, coordination, and division of labour and expertise not only in finding vaccines for global viruses, such as COVID-19, but also comprehensively prepare for a better post-COVID-19 world. Through strategic management, with the right attitudes, governments, international agencies, and all others concerned (stakeholders) can act together in a coordinated manner to identify and tap the world's strengths and resources, more equitably and effectively, and assign roles and responsibilities which reinforce each other at both the national and global levels.

There is no shortage of governments, organizations (including numerous moribund ones) and even resources in this world to fight COVID-19.[21] However, there is a very serious deficit in coordination, common interest, and common understanding in waging the battles against COVID-19 with the same efficiency that we have fought world wars before. Just as ESG is more than shareholders, our world must be more than countries or nations; as ESG is moving company's concerns and responsibilities from only shareholders to all stakeholders, governments must also move their concerns beyond rhetoric from countries or nations to the global village.

Hence, it cannot be overemphasized that, in the interest of efficiency, governments must now come together and function as one company, owned by one shareholder (humanity). Each continent and nation must be seen as a department in the company. The UN agencies, Bretton Woods institutions, central banks, ministries of agriculture, security forces and intelligence agencies, research agencies, power and energy enterprises, transportation and logistics authorities, and the world's greatest thinkers and business executives are needed to design, set up, and manage an effectively functioning coronavirus-resilient international system, which is over and above all existing international organizations, including the UN. These existing international organizations are commonly restricted in the

[21] Some of these moribund organizations are struggling to survive financially whilst others have significant resources. The world may be better off identifying and shutting them down.

pursuit of their objectives by the self-interest of the most powerful and bureaucracy of its machinery. ESG and the COVID-19 pandemic calls for pressures not only on companies, but on governments, especially the most powerful ones. The most powerful nations/governments must lead the way, just as they have led the world in two world, and other major, wars. They must be ready for greater sacrifice for the global good and such readiness should be inspirational to African governments and the private sector in Africa, in particular, and to the rest of the world. Historical and contemporary facts have proven that the most powerful countries (which also happen to be the developed only or mostly) can and do change the world.[22]

Calling for governments to think like companies, means the starting point is to set up a "One World Super Board" (no mundane names like organization, agency, commission, etc.). There is urgent need to start the process outside the orbit of the traditional international institutions (e.g. UN, Bretton Woods, G7, etc.), although the Board can draw expertise and support from the latter. The reason is simple and already stated—it is almost common knowledge that these organizations largely represent the interests of the developed/most powerful countries. The present global status quo is like the village dancing parties which I used to witness as a child. The drummers determined the tunes, the dancers danced and compete with each other on the floor and the admirers/spectators threw in all their money, all of which went to the drummers. The drummers are the developed/most powerful countries, the dancers and spectators are the rest of the world.

Business, academia, and civil society must be prominent, if not the most prominent component of the super board, and it must get the full backing of all the powers that matter at the highest levels in each country and in existing international institutions. It must have representation from every continent and function like the board of a private company with no restrictions, except what hurts the global interest. It must command respect from all stakeholders and have access to decision-makers in every country and the resources, mandate and ability to enforce its decisions. It must be headed by a proactive, objective, independent-minded, self-driven chief executive officer, with a solid track record in business, civil society, or the military.

[22] China and Russia are among the world's most powerful countries, but it may be arguable whether to count them among the developed countries.

Potentially, one of the biggest assets of this global "corporation" or superstructure presently is fear. Sweeney II (2020) remarked that fear is indeed a fuel: "The path to a fulfilling life is not to avoid fear but to recognize it, understand it, harness it, and unleash its power".[23] The universal fear that the coronavirus has induced through its indiscriminate invasion has made it a common threat which is creating the potential for a more caring world for all, but the reality is still elusive to a very large extent. To turn potential to reality, governments must not only be advocates of ESG but executors of a globalized ESG themselves, the most powerful ones in the lead, the rest following in the global village; this can lead a unified "people-centred development" programme at the global level. ESG is about caring for people at the company level, about capital caring for people, a globalized ESG is more about politics caring for people in Africa and the rest of the world.

To be successful, the superstructure must not be based on conferences of typically talk, talk, talk, and national or regional interest. It must be able to use existing economic and political structures or institutions to its advantage and not vice versa. National and regional interests must be subjected to the global interest(s), because we can only succeed to create a better and truly sustainable world through coordinated global action through strategic management with the global interest(s) in sight. As the saying goes, all for one and one for all. The world can do it but those with the powers to do so must be ready to lead with greater sincerity.

4.7 COVID-19, ESG and Africa: What Can Be Done by Companies for the Health Sectors?

In Chapter 2, we make a very elaborate and essential case for ESG in Africa. The relevance of ESG was first argued environmentally through various aspects of Africa's climate challenge that we exposed. Secondly, we do so socially, by covering the importance of addressing the needs of the women, youth, children, education and health sectors, rural areas, social protection in general, micro, small, and medium enterprises (MSMEs), and reversing the resource curse. Finally, in governance we called for the prioritization of corporate governance especially for local companies.

[23] Patrick J. Sweeney II, *Fear Is Fuel: The Surprising Power to Help You Find Purpose, Passion, and Performance*, Rowman & Littlefield Publishers, 2020, inside cover?

As this chapter is focussing on COVID-19 in Africa, consequently we address the question of what private companies can and should do to address the challenges of the health sectors specifically in the continent. This is in the interests of the private sector(s) in Africa as well as those of the larger societies. Therefore, our attention here will be on the Social (S) category of ESG which encompasses health and safety for company employees first and, by extension, can and should be made to reach the communities too. As we have portrayed the sorry state of the African health sectors in Chapter 2, we do not repeat those details here, except when emphasis is required or additional data is presented.

Our recommendations for corporate social actions follow. These draw upon our practical experience in providing health facilities and services, based on CSR, to communities in Africa, in particular from August 2006 to August 2011 in Mauritania. During this period, we built the country's first emergency ward, a number of community clinics, provided medical equipment and training, and carried out several mobile medical clinics in nomadic settlements in the Sahara desert. During one of these trips in the desert, on the 15th of January 2010, our team delivered a baby girl, whose parents named her after our company (Tasiast)! The lessons we have learned in Mauritania and since then in a number of African countries clearly demonstrate that effective coordination between the companies, communities, and governments is essential in any African country in order to encourage companies to fund CSR projects in traditionally public domains, such as the health sector, and to make these projects sustainable.[24]

[24] The first author of this book was Vice President of Tasiast Mauritania Ltd., a mining subsidiary of Canadian companies, Red Back Mining, then Kinross Gold. As Vice President during one of the biggest discoveries of gold in Africa in the decade of the 2000s, he was directly responsible for community and government relations, business development, and, over a number of years, for environmental oversight in the company. In addition to this, both authors have been involved in community health projects in East, North, and West Africa through their own foundation, Heeno International, since 2011. Our experience proves two things: (1) Financing—the necessity for companies to be involved in building better health facilities and services in Africa through private funding and the management of such funding and (2) sustainability—the abilities or willingness of the governments and communities to maintain these facilities and services at desirable standards over a very long period of time.

4.7.1 Existing Medical Facilities

It is public knowledge that throughout Africa, existing medical facilities, especially outside the urban areas, generally exist in dismal states. Even in urban areas, few of these facilities, for very small percentages of the populations who can afford them, meet international standards. The unappealing facilities dent the abilities and morale of smart and dedicated personnel working there, resulting in substandard patient care. Therefore, companies can provide equipment to the hospitals, clinics, and other medical facilities and help renovate them as part of their ESG programmes.

4.7.2 Working Medical Personnel

Medical personnel in Africa can be grouped into three categories—those trained in recognized schools in developed countries at international standards and those trained at institutions at home or in other African or developing countries. Many (but not all) of the latter institutions lack the resources to train at international standards. Companies can help, through their ESG programmes, by inviting medical experts from different parts of the world for short-term refresher or upgrading courses for African doctors, nurses, and other health workers, especially those without the required international-level professional exposure.

4.7.3 New Medical Facilities

Existing medical facilities in Africa are not only generally substandard (as noted above) but their numbers are also grossly insufficient for the growing populations. Furthermore, they are very much unevenly distributed in favour of the urban areas, especially in and near the capital cities. Therefore, new, better, and more evenly distributed facilities are needed to expand the access of the populations to medical services. Companies can allocate part of their CSR budgets, under ESG, for achieving this objective.

4.7.4 Additional Medical Personnel Through Academic Institutions

Related to the need for new medical facilities, there is also the need to train more medical personnel to attain international academic standards.

This calls for better equipment, more and better qualified teaching staffs, and more appropriate curricula in academic institutions. Companies can allocate part of their CSR budgets, under ESG, for this purpose. As is now common practice in some of the developed countries, especially in the United States, private companies can help attract the best brains to poorly funded academic institutions in Africa by providing endowment chairs and research grants for faculty and scholarships for the best students.

4.8 Need for Clarity of Roles and Effective Coordination

In all the above, the role of companies should be limited to providing and controlling the funding necessary and making sure that the required standards are met in executing the programmes. Companies have their core businesses and would not like to be distracted from these to become "mini-governments" in Africa, as already emphasized in Chapter 2. The post-implementation phases of these projects, including in particular, their sustainability must lie with the governments and communities. Companies are not meant for these and cannot do so and forcing them to do these under normal circumstances would even discourage them for funding voluntary Social programmes.

Furthermore, the priority for companies should be the communities in which they operate and from where they benefit. Often in Africa, companies make dividends, taxes, royalties, and other payments to governments which are not spent fairly in the countries by the governments concerned. Common victims are communities in which natural resources are discovered and those in close proximity to such discoveries. Very little (if any) of the payments to the governments typically reach these source communities and instead they are left to bear any negative consequences of the extractive activities of the companies. Such unjust treatment results create the potential of conflicts between the communities and the companies because the former see the latter as exploiting their resources without any benefits to them. The communities find it easier to blame the companies than their own governments. Sometimes the governments benefit from such attitudes of the communities and leave the companies on their own to swim or sink. At other times (as is common in the past when governments and businesses had a freer hand in this world) the governments take the sides of companies with brutal and fatal force. The hanging of Ken Saro-Wiwa and the Ogoni activists in 1995 is one of the best examples.

These activists spoke out forcefully and rose up against the Nigerian military regime, led by Sani Abacha from 1993 to 1998, and the Anglo-Dutch petroleum giant, Royal Dutch/Shell.[25] Because of such experience, therefore, the CSR projects of companies have to prioritize their immediate environments to make their positive presence felt, first and foremost, by the inhabitants of the communities where they exist and directly benefit from.

4.9 Dumping Toxic Waste and Deadly Sub-Standard Medication

In October 2022, 70 Gambian children died as the result of adulterated Indian syrups. Since the 1980s, we have been interested in toxic waste dumping in Africa by foreign companies and have written about it. In 1991, the World Bank Vice President attempted to suggest that Africa could be the right place for such dumping on economic grounds. This kind of critical issues highlight the importance of ESG and the responsibilities of businesses, governments, and international institutions.[26]

[25] For more information, for instance: https://www.britannica.com/biography/Ken-Saro-Wiwa; https://www.ihrb.org/other/on-the-life-and-legacy-of-ken-saro-wiwa-iii and https://www.sahistory.org.za/people/kenule-beeson-saro-wiwa, last visited December 2, 2022.

[26] https://www.reuters.com/world/africa/gambia-tightens-rules-indian-drugs-after-cough-syrup-deaths-letter-2023-06-20/#:~:text=At%20least%2070%20children%2C%20most,adulterated%20cough%20syrups%20from%20India; https://www.harvardmagazine.com/2001/05/toxic-memo.html

CHAPTER 5

Designing and Reporting ESG Disclosures

Abstract In this chapter, we take the reader through the practical design and application of ESG reporting. We explain concepts that are essential and opportunities for sustainability that can be utilized in the process. We explore the list of reporting formats and rating agencies that dominate the international ESG reporting landscape. We also present actual ESG/sustainability reports of different organizations and end up with one that we have designed and applied to a unique and operating project of empowerment and inclusion for the benefit of persons of special needs in Africa.

Keywords Sustainability · Circular economy · Sustainable finance · Materiality analysis

...it is important to recognise that each country is different and therefore has different needs, challenges and capabilities. The needs of developed and developing countries, established and emerging economies, are very different. Countries with citizens in extreme poverty have contrasting challenges to first world nations, including where to invest precious resources and brainpower, and focus policy and social mobilization efforts, which are particular to each state. Therefore, African countries should not simply copy and paste ESG measures to address their needs, instead should be able to assess their

K. N. M. Sonko and M. Sonko, *Demystifying Environmental, Social and Governance (ESG)*, Palgrave Studies in Impact Finance,
https://doi.org/10.1007/978-3-031-35867-8_5

circumstances, economies and societies to determine what they should consider material to their urgent challenges, whilst also recognizing the global threats which they share with the rest of the world.[1]

5.1 What to Take Into Consideration

In this chapter we look at the practical aspects of ESG by outlining the processes of designing and reporting or disclosing ESG measures. The design of ESG frameworks for reporting or disclosing can be easy or difficult, brief or extensive, be unique to each company or can be based on a standardized framework selected based on preference for one reason or another. The focus of the chapter is on companies/corporations, whether listed or not, state-owned or private, foreign or locally owned, that operate within the formal sector of the economy. In Africa, where few countries have fully fledged stock exchanges, state-owned corporations are common in all countries and foreign corporations tend to be dominant in industries that require large capital outlays and advanced technologies, such as the extractive industries. Therefore, companies or corporations in this chapter refer to all businesses incorporated in the formal sector, except those that are too small in terms of revenue, impact (especially on the environment), and number of employees. The latter are commonly sole proprietorships and family enterprises that operate in the informal sector without official registrations.

Generally, in the continent, compliance with ESG issues tend to be expected mainly of the private, instead of public, sector companies and foreign (especially big), rather than locally owned, companies. On the contrary, we believe that it is essential to not limit ESG reporting to only a certain sector but to look at all potential candidates for ESG disclosures in the formal sector, public and private, foreign and locally owned. In fact, we sometimes use the terminologies of "institutions" and "organizations" to broaden our perspectives beyond private or profit-making companies to not-for-profit structures, the activities of which make them potential ESG candidates, liable for all or certain ESG issues.

[1] Kgalema Motlanthe, "Remarks at the First High-level Webinar on ESG and Stakeholder Capitalism in Africa," Virtual Conference organized by Heeno International and partners on 28 January 2022.

According to the London Stock Exchange Group (LSEG), there are certain "priorities" to take into account when designing ESG Reporting.[2] We take some of these priorities and include additional ones of our own, which we believe are equally relevant in the African context. The discussions here are for investors, managers, and others interested in designing ESG disclosures, knowing the commonly referred to frameworks, and gaining an idea of the various ESG rating agencies and their indexes.

The first is *strategic relevance* for listed companies and those depending on third party investors with ESG concerns. Such investors want to understand how issuers or those seeking capital respond or propose to respond to long-term and macroeconomic trends such as climate change, taxation, demographic, and technological changes as well as political developments.

Secondly, *materiality*, is a fundamental concept in ESG reporting, which resonates throughout the system. Issuers and managers should decide which ESG issues they see as most relevant or material to their businesses. This can be determined formally in accordance with the Sustainability Accounting Standards Board (SASB) Materiality Map®, listing material items and the levels of their significance in a materiality map. This map is based more on an inward-out perspective; i.e. the impact of the external environment on the company.[3]

The concept of materiality originally derives from financial reporting. The principle is that audit practices must filter and guarantee information in a manner that ensures true, fair, and useful representation of the company's financial situation, in support of capital protection, risk management, and decision-making (Edgley, 2014; Messier, Martinov-Bennie, & Eilifsen, 2005). More recently, the concept of materiality has been adapted to sustainability reporting. In this area, the principle is that information must be filtered and guaranteed in support of those company activities and decisions impacting society and the environment, and affecting present and future stakeholder needs (Hahn & Kühnen, 2013). This may be referred to as the "inside-out or in-out" perspective and distinguished from the SASB's more "outside-in or out-in" approach,

[2] LSEG, *Revealing the full picture. Your guide to ESG reporting Guidance for issuers on the integration of ESG into investor reporting and communication*. London, 2020; the reader may also refer to Steward Redqueen, European Bank for Reconstruction and Development (EBRD) and the Warsaw Stock Exchange (WSE/GPW), May 2021.

[3] For more information on the map, the interested reader should refer to: https://www.sasb.org/standards/materiality-map/.

for example. In the Global Reporting Initiative (GRI) guidelines, a widely accepted standard for sustainability reporting (Calace, 2016), "material aspects" are those that reflect the company's significant economic, environmental, and social impacts, or those which significantly influence stakeholders' assessments and decisions (GRI, 2016, 2013a, 2013b).

In this book materiality analysis is defined as the process through which a company identifies, selects, prioritizes, and reviews what is material to the company and its stakeholders, and thus merits inclusion in the ESG or sustainability reports (GRI, 2013a, 2013b). For companies with the capacity to do materiality analysis, the process should be conceived as systematic and rigorous, contributing to the identification of significant stakeholder-oriented metrics of corporate strategy. Such metrics can assist the company in enhancing awareness and visibility of its sustainability performance, by improving the readability and quality of reporting for different actors (Carpejani, Pinheiro de Lima, Gouvea da Costa, Machado, & Da Veiga, 2017; Russo-Spena, Tregua, & De Chiara, 2016). The metrics can assist in benchmarking, initiatives for stakeholder engagement (Forstater et al., 2006), monitoring performance, and identifying improvement opportunities.

According to Font et al. (2016), materiality analysis can in fact serve in redirecting corporate strategy and the reporting towards greater inclusion of stakeholder needs. The latter can in turn facilitate the creation of shared value, since it identifies the issues with the highest potential to benefit both the company and its stakeholders.

When sustainability reports deal with material issues, they reveal corporate strategy as a driver of innovation and economic growth, disclosing the inputs for managing the future (Bowers, 2010).[4] Through materiality analysis, companies can improve their accountability to stakeholders and make sustainability efforts more effective. Indeed, a materiality analysis approach to sustainability reporting (based upon an acceptable framework/method) allows dealing with fundamental theoretical concerns such as subjectivity in assessment and consistency of judgements; complexity and uncertainty of sustainability issues; the need to meet different stakeholders' expectations; balancing between standardized assessment procedures (that ensures comparability of reporting) and the

[4] Bowers, T., "From Image to Economic Value: A Genre Analysis of Sustainability Reporting", *Corporate Communications: An International Journal*, Vol. 15, No. 3 (2010): 249–262. https://doi.org/10.1108/13563281011068113.

need to take into account the specificity of different contexts (Ascough, Maier, Ravalico, & Strudley, 2008; Greenwood, 2007; Manetti, 2011; Zhou, 2011). The process of assessing materiality requires companies to consider the strategic value of the opportunities and risks related to the various aspects of sustainability. This means considering how much the company's impacts are crucial to the company's performance, to its ability to grow and obtain competitive advantage (GRI, 2013b). This is an advanced form of materiality because of the process and detail required, unlike the simple process of materiality in which a company objectively or subjectively decides which issues are material and worth reporting.

Returning to the priorities, thirdly, a manager overseeing ESG reporting or all managers in an company should articulate the ***key performance indicators (KPIs)*** or metrics in the areas that are identified (Table 5.1). Key performance indicators provide concrete outputs of a company's ESG issues through performance measurements which enable it to determine success or failure in its goals. To be useful, KPIs must be quantifiable to enable managers to see changes in the activities targeting the ESG goals. The KPIs must also reflect and be directly relevant to the issues identified, as shown in the examples in Table 5.2. A combination of KPIs with SMART (Specific, Measurable, Achievable, Realistic and Time-bound) goals make the ESG process more practical and useful.[5] The African Pestle Analysis (APA) in Chapter 3 makes the combination of both possible in its framework. Furthermore, the APA can help decision-makers avoid the dangers of broad and ineffective KPIs by providing the organizational basis through the identification of the relevant challenges. Performance Square, which is a part of the APA, helps to further narrow down activities into targetable, quantifiable, assignable, and manageable objectives, processes, and pursuits.

[5] For very simple illustrations of goals and KPIs, see for instance, Tableau, "What Are SMART Goals? Definition, Best Practices, And How To Make Your Own," https://www.tableau.com/learn/articles/smart-goals-criteria, retrieved April 24, 2023. For in-depth or scholarly analysis the interested reader may refer to Robert S. Rubin, "Will the Real SMART Goals Please Stand Up?" https://citeseerx.ist.psu.edu/viewdoc/download?doi=10.1.1.523.6999&rep=rep1&type=pdf, retrieved April 25, 2023; Doran, G. T. (1981), "There's a S.M.A.R.T. Way to Write Management's Goals and Objectives," *Management Review*, 70 (11): 35–36; and Peter Drucker, *The Practice of Management*, New York: Harper, 1954; or later editions—Heinemann, London, 1955; revised edition, Butterworth-Heinemann, 2007; and Queensland Government, "Short and Long-Term Goals and KPIs," https://www.business.qld.gov.au/running-business/planning/goals-kpi, last visited April 24, 2023.

Table 5.1 Examples of KPIs in ESG reporting

Sector	*KPI*	*Specification*
Home Improvement Retailers	Energy Efficiency	Energy consumption, total
	GHG Emissions	GHG Emissions, total (scope I, II)
	Waste Scope I	Total Waste in tonnes
	Supply Chain	Total number of suppliers
Gold Mining	Energy Efficiency	Energy consumption, total
	Corruption	Percentage of revenues in regions with Transparency International corruption index below 6.0
	Water Consumption	Water consumption in m^3
	Training and Qualification	Average expenses on training per FTE p.a
Farming and Fishing	Energy Efficiency	Energy consumption, total
	Corruption	Percentage of revenues in regions with Transparency International corruption index below 6.0
	Water Consumption	Water consumption in m^3
	Training and Qualification	Average expenses on training per FTE p.a
Furnishings	Customer Satisfaction	Percentage of total customers surveyed comprising satisfied customers
	Waste Scope III	Hazardous waste total in tonnes total
Home Construction	Energy Efficiency	Energy consumption, total
	GHG Emissions	GHG Emissions, total (scope I, II)
Consumer Electronics	Energy Efficiency	Energy consumption, total
	GHG Emissions	GHG Emissions, total (scope I, II)

Source Authors, based on selected KPIs for ESG in Different Sectors from European Federation of Financial Analysts Societies (EFFAS) and Society of Investment Professionals in Germany (DVFA), A Guideline for the integration of ESG into Financial Analysis and Corporate Valuation, Version 3.0, 2010, Frankfurt am Main, 2010

Table 5.2 Summary of the contents of ESG/sustainability reports

Company	*Company*	*Company*	*Company*	*Company*	*Company*
Perseus Mining, (Gold, Australia) *Sustainable Development Report 2021*	MTN Group (Telecomms, South Africa) *Sustainability Report 2021*	DEWA (Utilities, UAE) *Sustainability Report 2020*	RHB (Bank, Malaysia) *Sustainability Report 2021*	A.P. Moller (Fund management, Denmark) *Capital ESG Report for 2020*	Dubai Government (UAE) *Expo2020 Dubai Sustainability Report*
Table of Contents	Table of Contents	Table of Contents	Table of Contents	Table of Contents	Table of Contents
Letter from Group General Manager Sustainability	Doing for Tomorrow, Today	About DEWA	Introduction	Executive Summary	About Expo2020 Dubai
Highlights	Doing for our Planet	Sustainable Development	Sustainable & Responsible Banking	Our Approach to ESG & Impact	Foundations for a Sustainable Expo
About Perseus	Doing for Society	Energy	Embedding Good Practices	ESG Framework	The Expo Legacy
Our Sustainability Performance	Doing it Right	Water	Enriching & Empowering Communities	Fund Positive Social Impact	Expo 2020 & the SDGs
Our Contribution to the UN SDGs	Doing for Growth	Climate Change & the Environment	Sustainability Framework	Exclusion List	GRI Content Index & Topic Boundaries
Our Approach to Sustainability	ESG Indicators Performance	Employees	Sustainability Governance	22 pages	238 pages
Our Material Issues	78 pages	Customers	Sustainability Principles		
Sustainability Data Tables		Community	Engaging with Stakeholders		
142 pages		Context Index	Assessing our Material Matters		

(continued)

Table 5.2 (continued)

Company	*Company*	*Company*	*Company*	*Company*	*Company*
		140 pages	Alignment to the UN SDGs Appendices 124 pages		

Source Authors, based on the reports of the organizations cited

Fourthly, ***benchmarking***. In addition to materiality analysis, companies usually perform benchmarking assessments to compare themselves to their competitors. Benchmarking began in the late 1970s as a philosophy that seeks the best practices that guide a company to maximize its performance. The initial milestone was the study by Xerox that sought to compare its operations with those of its competitors (Elmuti and Kathawala 1997). Organizations began learning what and how leading companies do to reach the top positions. The analysis of the processes, regardless of what they are, offers the opportunity to evaluate the quality of the processes and learn lessons that can be adapted to the specific reality of another business or activity (Seibel 2004).

Fifthly the ***data*** used is very important.[6] When using ESG data to inform capital allocation and investment decision-making, investors want ESG information to be complete, consistent, reliable, comparable, and clear. Such investment grade data is essential whether the company is listed or not, if it intends to publish its ESG disclosures. Reliable data is needed not only for the investment decisions of a company, but for all variables that require numerical analysis, such as population, infrastructure, employment, and income levels, in a catchment area for community engagement by a company.

Having dealt at length with rating agencies, we now turn, seventhly, to *global frameworks* in ESG reporting of which there are also many to choose from, developed by different organizations too. Among them, the most sought after by investors are those of GRI, the International Integrated Reporting Council (IIRC), SASB, the UN Global Compact, the CDP (formerly the Carbon Disclosure Project), the Climate Disclosure Standards Board (CDSB), and the Financial Stability Board (FSB)'s Task Force on Climate-Related Financial Disclosures (TCFD). The evolving metrics of the World Economic Forum, IBC, and the Big Five, discussed at length in Chapter 1, are now being promoted by their proponents.

Before we move on to the eighth factor to take into consideration, the most widely used and recognized frameworks are summarized below for the benefit of the reader. More information can be obtained on them from the respective websites.

[6] On this point the reader may also refer to World Business Council for Sustainable Development (WBCSD) & FSR—Danish Auditors, "Guidance on improving the quality of ESG information for decision-making. Developing a roadmap for companies," June 2019.

5.1.1 The Global Reporting Initiative (GRI)

This international independent body helps businesses, governments, and other organizations understand, develop, and communicate sustainability metrics. The guidelines, also known as "GRI standards" can be downloaded freely on their website. GRI relies on voluntary disclosure. Managers have the choice to select topics for their ESG reports that they consider most relevant or material to the company and share these with the shareholders. GRI is the most well-known reporting framework.

5.1.2 The Sustainability Accounting Standards Board (SASB)

This non-profit organization has developed global standards for identifying, managing, and communicating financially material sustainability information to investors. These standards are explained through a materiality (discussed in Section 5.1) map and contain a set of 77 industry-specific metrics. SASB is a great choice when managers want to determine materiality in an objective manner and communicate the value they create in investor language. Besides, managers can use it in conjunction with other frameworks. Many companies use SASB along with GRI, for example, to achieve double materiality, which enables the company to capture its impact on the external environment and vice versa.[7] The APA can also make this possible in its strategic and comprehensive framework.

5.1.3 International Integrated Reporting Council (IIRC)

The IIRC first published the International Integrated Reporting Framework in 2013 and revised it extensively in 2021. They defined integrated reporting as "a process founded on integrated thinking that results in a periodic integrated report by an organization about value creation over

[7] For an introduction to double materiality and why it is important, the reader may refer to Matthias Täger, "'Double Materiality': What Is It and Why Does It Matter?" Commentary, the London School of Economics and Political Science, Grantham Research Institute, 21 April 2021, retrieved April 25, 2023.

time and related communications regarding aspects of value creation". Its aims are to:

- Improve the quality of information to enable a more efficient and productive capital allocation by providers of financial capital.
- Promote a more cohesive or integrative and efficient approach to corporate reporting, drawing on different strands and communicating the full range of materiality factors that affect an organization's ability to create value in the short, medium, and long terms.
- Enhance understanding, accountability, and stewardship with regard to different types of capital (financial, manufacturing, intellectual, human), and promote understanding of their independencies.
- Apply integrated thinking in reporting for businesses of any size as well as public sector and not-for-profit organizations.

The IIRC's framework is often used together with SASB standards.

5.1.4 The Workforce Disclosure Initiative (WDI)

This is an investor collective created to help companies to better communicate labour practices to stakeholders in an efficient way. It aims to improve corporate transparency and accountability with regard to labour and job creation internationally and to furnish investors and companies with comparable comprehensive data.

The WDI investor coalition is made up of 68 institutions, with $10 trillion in assets under management. Through their annual survey and engagement programme the WDI generate data on workforce practices, which their members use for investment analysis and gaining insights on how to address urgent labour matters. The WDI reported that 173 multinational corporations participated in their programme, representing an increase of 23 per cent in 2020.[8]

[8] For more on WDI, visit their website: https://shareaction.org/investor-initiatives/workforce-disclosure-initiative.

5.1.5 *The Task Force on Climate-Related Financial Disclosures (TCFD)*

An initiative of the Financial Stability Board, the task force was created to improve and increase reporting of climate-related financial risks. The TCFD prescriptions for climate-related financial reporting can be found in their recommendations outlined in a report along with in-depth tutorials on how to adhere to their standards. They can be applied to a broad range of companies in different activities, sectors, and jurisdictions. The recommendations fall into four areas in order to capture the main features of functioning institutions. The themes are governance, strategy, risk management, and metrics and targets. The United Kingdom is considering implementing the TCFD requirements in its legislation. Other jurisdictions, including South Africa, Canada, the European Union, Japan, and Singapore, already have reporting requirements aligned to TCFD standards.

5.1.6 *The Climate Disclosure Standards Board (CDSB)*

This is an international consortium of NGOs that were set up to help organizations integrate information related to climate change in their financial reporting. The CDSB has developed its own framework which is currently used by 374 companies across 32 countries. On 31 January 2022, the CDSB was consolidated into the International Financial Reporting Standards (IFRS) Foundation in order to strengthen the work of the International Sustainability Standards Board (ISSB), a new entity. Following this decision, the CDSB has ceased to provide any guidance or other work on sustainability disclosures and it is likely that the IFRS will now be the one to issue the Sustainability Disclosure Standards in due course.

Sixthly, there are various ESG *rating agencies* which are organizations that examine a company's environmental, social, and corporate governance policies to determine its ESG performance. Investors can use an ESG rating agency's reports to assess how firms of interest to them score on ESG criteria, enabling the investors to better determine which ones to invest in. Many universities, pension funds, social trading funds, and other institutional investors searching for long-term financial stability consider ESG ratings when making investment decisions.[9]

An ESG rating agency is usually hired by an interested company to perform an in-depth analysis of its practices, and then help it develop better policies. ESG ratings providers are also becoming increasingly popular among companies who aim for good publicity (and increased revenue)—something which can be achieved with positive ESG developments.

In the area of greenhouse gas emissions, for instance, the International Organization for Standardization has announced a new initiative to help businesses measure and improve their sustainability performance through reduction of their carbon footprints. With certain firms doing more harm than good in this area, rating agencies can hold them accountable for their actions.

The main rating agencies and their indexes include:

5.1.7 Sustainalytics ESG Risk Ratings

Sustainalytics is an ESG rating and data supplier that provides ESG ratings on 20,000 companies in 172 countries. They rate 40,000 companies worldwide. Sustainalytics is a subsidiary of Morningstar, one of the largest stock market data providers in the world. The ESG ratings from Sustainalytics measure environmental, social, and corporate governance performance of companies on a global scale. They cover about 13,000 international equities across all regions worldwide.

ESG ratings are or should be based on both quantitative ESG data and qualitative analysis.[10]

[9] In a study by Christensen, Serafeim and Sikochi (2021) it was found that the number of these agencies has increased dramatically and there are disagreements among them on how to rate companies and therefore the outcomes. See Dane Christensen, George Serafeim and Anywhere Sikochi, "Why is Corporate Virtue in the Eye of The Beholder? The Case of ESG Ratings," Harvard Business School, Working Paper 20-084, 2021. On impact and measurement challenges see also Sara Lovisolo, "Global Environmental, Social and Governance (ESG) Capital Allocation Strategies Between Impact Ambitions and Measurement Challenges," IAI Papers 21/35, July 2021; and Elizabeth Pollman, "The Making and Meaning of ESG," University of Pennsylvania and ECGI, Law Working Paper N° 659/2022 September 2022.

[10] For an example of such scores the reader may refer to Refinitiv, "Environmental, Social and Governance Scores from Refinitiv," LSEG, May 2022, https://drive.google.com/file/d/1dRi3iAETlHrU_d-osAWqzMoF3IyMZl0H/view, accessed April 12, 2023.

ESG Scores cover several different areas including governance, environmental impact, social contribution, and financial performance to provide a holistic view of the ESG profile of companies. The level of consistency with the information provided by Sustainalytics makes it one of the top firms in reporting and data for ESG.

5.1.8 MSCI ESG Ratings

MSCI ESG Ratings are created by MSCI ESG Research, one of the largest rating agencies. These ESG ratings are released for 14,000 different equity and fixed income issuers. The MSCI ESG Ratings are generally known to be one of the industry leaders in publishing scores and ratings for ESG practising companies. For example, the MSCI USA ESG Leaders Index tracks the performance of U.S. companies. It covers large and mid-capitalization American stocks. The BMO MSCI USA ESG Index ETF focuses on those in this index that have higher MSCI ESG ratings than the others. The ETF uses this index to select and invest in the constituent securities according to their performances in the Index.

5.1.9 Bloomberg ESG Disclosures Scores

Bloomberg ESG Disclosure Scores is an ESG data system that provides ESG information for more than 11,800 companies in over 100 countries. Their ESG data includes topics such as climate change, human capital, and shareholders' rights. The ESG Disclosure Scores rank companies on basis of their levels of ESG disclosure and include key sustainability topics.

5.1.10 FTSE Russell's ESG Ratings

FTSE Russell has consolidated indicators from multiple global standards and developed an Exposure framework. The framework identifies indicator applicability for various companies depending on their industrial sectors and geographic locations.

The ESG Ratings by FTSE Russel are an assessment system of a company's ESG performance. They consist of more than 7200 securities from 47 countries and are based on a methodical analysis of performance at a company level. The ratings are a way of comparing and analyzing the ESG performance of issuers. Six ESG areas are covered:

- Corporate Governance
- Environmental Policy
- Social Policy
- Labour Practices
- Supply Chain Policy
- Country of Origin (a proxy for economic development).

It is important to note that these ratings are companies that are listed in the FTSE Global Equity Index Series and other large and liquid stocks. Therefore, they give a better insight into mid to large-cap issuers and not suitable for small-cap or micro-cap security issuers.

5.1.11 Institutional Shareholder Services Ratings and Rankings

Institutional Shareholder Services (ISS), majority-owned by Deutsche Bourse Group, provides ratings and data analysis on companies, countries, and funds, across a full range of sustainable investment issues. These issues cover climate change, human rights, supply chain management, labour practices, and even corruption and controversial weapons.

ISS ESG ratings can help investors in determining ESG compliance and risks faced by companies. The ESG performance is assessed on the basis of company-specific key performance indicators (KPIs) in the areas that are identified above.

5.1.12 S&P Global ESG Scores

Standard & Poor's Global is one of the largest companies that provide data analytics and reporting related to companies around the world. S&P Global argues that many ESG scorecards take a top-down approach, following the idea that ESG scores should be decided by a higher entity, in other words top management. On the contrary, they point out that their ESG Scores are different from other ESG scoring systems because they are based on a bottom-up approach. The scores can help provide more information on a company's ESG performance, such as how well it works to minimize ESG risks. The ESG scores for over 11,500 companies are calculated based on answers from companies in S&P's ESG questionnaire and/or publicly available data collected by S&P. The ESG questionnaire has more than 450 questions.

They maintain that their system's goal is to take the ESG Scorecard directly to the industry level, where the ESG behaviours of companies can be controlled. S&P Global ESG Scores takes a more analytical look at how well companies perform in areas such as environmental practices and employee relations. Therefore, it enables both investors and consumers to better understand ESG risks and opportunities. The ESG ratings for each company are published on S&P's ESG website.

5.1.13 CDP Climate, Water, and Forest Scores

A not-for-profit organization, CDP provides environmental data, research, and tools for investors. It helps them identify funds that invest in companies that are successful at addressing material concerns connected with climate change, water security, and deforestation. Climetrics rates almost 20,000 funds and publish them freely. The CDP ESG Rating is a unique rating that identifies what CDP considers the best ESG-integrated investment funds. It includes all types of fund products (i.e. mutual funds, ETFs, and separate account portfolios) and can be used by investors to identify the ESG funds in my portfolio.

5.1.14 Moody's ESG Solutions Group

Moody's ESG Solutions Group is a Moody's Corporation business unit. Moody's is well known as one of the largest credit rating agencies in the world. This segment of their business offers ESG ratings, analytics, sustainability ratings, and sustainable finance reviewer/certifier services, using their own data. The group now includes ESG assessor Vigeo Eiris as well as climate data company Four Twenty Seven, which were both acquired in 2019. Moody's ratings are designed for investors who want ESG-related information on companies across all industries, countries, and regions. The ratings include over 13,000 ESG assessments.

Next, the eighth priority to take into account in ESG reporting is the *reporting format*: companies can report ESG information in their annual reports, in a stand-alone sustainability report or in an integrated report. The choice of reporting formats may involve trade-offs between breadth and depth, between focusing on material issues and covering a wider horizon that addresses the relationship between ESG and business strategy.

Ninth, managers should know which relevant ***laws and regulations*** concern his/her company with regard to ESG. The volume of regulations, nationally and internationally, concerning ESG reporting has increased substantially in the last decade. According to the LSEG, "over 80% of the world's top economies by GDP in 2016 mandated ESG reporting in some form".[11] If regulators in different countries and regions set different reporting requirements and standards, this can be problematic for both issuers and investors operating across different jurisdictions. An ESG manager should know how to navigate through this complex network of requirements.[12]

Tenth, ESG managers, in particular, should be interested in ***green reporting***, which is part of sustainable finance.[13] This requires companies or organizations to report revenues or expenditures that have an environmental impact. Green reporting can be done through the standard environmental impact assessments at the beginning of a project on a case-by-case basis and/or be integrated in the continuous operations of companies and organizations. It involves determining the impacts of an organization's activities or projects on the environment in terms of climate change, mitigation, and broader concerns. The standard that is now being considered in some circles as the global gold standard is the

[11] Most of the mandatory reporting is in the environmental area where are many laws now across the world, although not equally enforced across countries.

[12] ESG is still going through an evolutionary phase with a lack of complete clarity on a number of issues, such as what should be mandatory and what should not be. However, much of what tends to be mandatory so far is in the environmental domain and increasingly community economic empowerment. The rest is mainly voluntary, peer-pressured or for public relations.

[13] For more on Green Finance the interested reader may refer to the Green Finance Platform, "Explore Green Finance", Geneva, 2021, https://www.greenfinanceplatform.org/page/explore-green-finance; Sean Fleming, "What Is Green Finance and Why Is It Important?" World Economic Forum, 9 November 2020, https://www.weforum.org/agenda/2020/11/what-is-green-finance/, last visited 14 March, 2022; *Journal of Sustainable Finance & Investment*, Special Issue: Sustainable Finance for Continued Entrepreneurship Growth in Developing Economies, Volume 12, Issue 1, 2022.

European Union's Taxonomy Regulation adopted in June 2020.[14] It aims to make the European Union (EU) climate neutral by 2050 and make the Union a global leader in sustainable finance. The taxonomy provides six objectives by which investments are to be assessed. These are climate change mitigation, climate change adaptation, the circular economy, pollution, effect on water, and biodiversity. In effect, these objectives are also criteria by which institutions can measure or determine their environmental impacts in order to meet the requirement of "Do No Significant Harm" (DNSH) environmentally.[15] The issue of adopting these same standards globally, especially in the developing countries, is a matter of great disagreement within the United Nations system and bilaterally between countries, especially the donors and recipient countries.

Eleventh, is about *information and communication*. If a company decides to engage in ESG reporting it should also think about how and to whom that information should be communicated. Unlisted companies should think about the government, NGOs, the media, and the general public, whilst listed companies should think about all of these as well as their investors and exchanges they are listed on.

Lastly, the *capacity* of a company to formulate and monitor ESG disclosures is extremely important and must be taken into account in shaping its ESG reporting. Generally speaking, MSMEs should not be expected to have the same capacities in implementing ESG programmes as large local companies, just as large local companies may not have the same capacities as huge multinational corporations. Therefore, the European Union's taxonomy may be considered almost ideal, but not for companies in Africa generally, especially locally owned ones without the necessary

[14] For more information on the Taxonomy see the European Commission (EC), Research and Innovation at the Heart of the EU Taxonomy, https://ec.europa.eu/info/sites/default/files/research_and_innovation/strategy_on_research_and_innovation/documents/ec_rtd_research-innovation-eu-taxonomy.pdf; and EC, Sustainable Finance and EU Taxonomy: Commission takes further steps to channel money towards sustainable activities, https://ec.europa.eu/commission/presscorner/detail/en/ip_21_1804, last visited July 17, 2022; and Holly Pettingale, Stéphane de Maupeou, and Peter Reilly, "EU Taxonomy and the Future of Reporting," Harvard Law School Forum on Corporate Governance, April 4, 2022.

[15] See for instance, OECD, *Developing Sustainable Finance Definitions and Taxonomies. A Brief for Policy Makers*, Paris, OECD, October 2020, available at https://www.oecd.org/environment/cc/developing-sustainable-finance-definitions-and-taxonomies-brief-for-policy-makers.pdf, last visited March 13, 2020.

resources, capacities, and regulatory support. Such companies cannot be expected to follow the same standards as in European countries. However, that must not mean that they should ignore the importance of the environment. Instead, it means that they should prioritize certain areas and pursue reasonable standards in line with their capacity constraints.[16] They should be provided the necessary support they need, especially in leveraging innovation and technology, through their governments and the governments of those countries advocating higher standards.

5.2 Summary of the Process of Designing an ESG Reporting Framework

From the discussion in this chapter up to this point, we can summarize the process of designing ESG disclosures/reporting as follows:

(1) Designing an ESG reporting/disclosure framework can be easy/simple or difficult, brief or extensive. As the Investment Company Institute pointed out, sustainable strategies may be incorporated in multiple ways.[17] An easy/simple framework would cover only a few main issues in the three ESG categories and might provide only a one-page report. At the end of the year, the relevant manager (s) can look at what has been achieved. Managers may wish to prepare instead a detailed reporting framework internally or with the help of external consultants, such as ESG Enterprise, whom the authors have a teaming agreement with. The report that is produced from such a framework can be a stand-alone, part of the company's annual report, or an integrated one, monitored monthly, yearly, or at other intervals. The report may be a simple word document, an excel file in computers, or in a computerized application provided by external consultants. The last type of report can be very detailed, technical, cover risks, and be related to overall business strategy (see for example, the Perseus, MTN group, and DEWA sustainability reports in Table 5.3).

[16] See Chapter 8 for prioritizing priorities for Africa as a whole.

[17] The Investment Company Institute, "Funds' Use of ESG Integration and Sustainable Investing Strategies: An Introduction," July 2020, file:///C:/Users/HP/Downloads/pdf_20_ppr_esg_integration.pdf, accessed April 12, 2023.

(2) An ESG report can be unique to each company or can be based on a standardized framework. Although there are ESG issues that are relevant to any company anywhere, the biggest shortcoming of standardized frameworks is that they tend to impose on, or assume, the same standards for each company in spite of geographic, economic, and cultural boundaries. Therefore, they contain an element of rigidity which may lead to resistance in certain regions, cultures, or countries and even in companies within countries. They also require expertise to be prepared and monitored.

(3) Because of point 2 above, materiality may be the most useful or essential process in ESG reporting as it enables every company to decide what issues are material to it or the environment in which it exists and to focus on them, within its capacity and constraints. As Motlanthe maintained, "it is important to recognize that each country is different and therefore has different needs, challenges and capabilities".[18] Using the flexibility provided by the concept of materiality, managers can decide what their companies and the natural and social environments require of them to report and what they can implement and monitor. As discussed earlier, linked to the concept of materiality are key performance indicators (KPIs), which put materiality in motion. In Table 5.2, we give examples of KPIs from different sectors, using the European Federation of Financial Analysts Societies (EFFAS) and Society of Investment Professionals in Germany (DVFA) reporting framework for ESG Version 3.0, 2010. KPIs may also be referred to as metrics (WEF, Chapter 1) or key sustainability indicators (MTN report in Table 5.3).

(4) For African companies that wish to be rated independently, particularly those that are listed and/or want to attract investments from Europe, North America, and from certain international financial institutions, it may be necessary to choose a rating agency from the list provided in this chapter. The agencies also choose companies for their databases on their own, based on publicly available and auditable data. Companies selected independently by rating agencies may provide additional, corrected (if necessary), or revised data to the agencies if they wish to. The African Development Bank

[18] Kgalema Motlanthe, ibid.?

Table 5.3 Disbursements, beneficiaries, and management of GSNEF in the experimental phase

Beneficiaries	*Male*	*Female*	*Total*
Entrepreneurial training (total number/persons)	175	234	409
Tailoring skills training (number/persons)	2	23	25
Microcredit (number/persons)	10	27	37*
Microcredit (total amount disbursed, Gambian dalasis)	130,000	284,000	414,000**
Other			
Advisory Board	2	2	4
Board Assistants	2	1	3

*19 directly financed by Heeno International Foundation and the balance by the Arab Bank for Economic Development in Africa (BADEA)
**D234,000 directly financed by Heeno International (HI) under GDITF
Source Authors, "The GSNEF Activity Report June 30, 2022," and training statement on 22 April, 2023

(AfDB), Africa's only AAA-rated financial institution, for instance, is rated by Vigeo Eiris.[19]

5.3 Opportunities That Can Make Reporting More Attractive

There are opportunities associated with ESG that reinforce sustainability and can make reporting or disclosures more attractive. Some are relatively new practices, some are not and all the main ones have been mentioned and defined in this and earlier chapters. At the top of the list in terms of popularity (and maybe utility too) are the circular economy and sustainable finance.

[19] One of the objectives of the Environmental and Social Safeguards Strengthening Action Plan (SSAP) 2020–2025 is to support capacity building in social governance (ESG) reporting Africa.

5.3.1 *The Circular Economy*

The circular economy is perceived as a crucial model for industrial economics to pursue sustainable development. It is seen "as a solution for harmonizing ambitions for economic growth and environmental protection" away from the wastage of the take–make–use–dispose linear economy. It is maintained that the "circular economy provides a reliable framework towards radically improving the present business model towards preventive and regenerative eco-industrial development as well as increased wellbeing based on recovered environmental integrity", despite the fact that the global pursuit for the circular economy needs further engagement (Eqqas, 2019; Rizos, 2017).

We put the circular economy at the top of the list of sustainability concepts not only because of its importance in industry but in all sectors of modern economies and in the life of every human being as a consumer and an economic agent/actor (see Chapter 8). Economic, social, political, and other organizational structures comprise individuals without whom they cannot exist and who define their characters and the impact on environment and others, including people on people themselves. Therefore, ESG measures can only have the desired full and comprehensive effects if individuals within and outside these structures fully influence or subscribe to the measures. In the case of the circular economy, companies may advocate or adopt environmental conservation measures (through recycling and other measures) as much as they want but if their employees, neighbourhood, and customers do not support the measures and continue to engage in conspicuous and wasteful consumption, their intentions, and even efforts, would have little or no effect beyond the papers that they may beautifully write on. For this reason, global and (not only) African efforts at sustainable development through the ingenuity embodied in the circular economy can only be successful if individuals change their lifestyles accordingly.[20] For this reason, population per se,

[20] For elaboration on this point, see Chapter 8 and scholarly studies of the circular economy as a concept and in practice see Erick Hungaro Arruda et al., "Circular economy: A brief literature review (2015–2020)," *Sustainable Operations and Computers*, Volume 2, 2021, pp. 79–86, retrieved April 25, 2023.

which we spend a lot of time and attention on may not be the problem because: (1) a small population can be more destruction to the environment than a bigger one, depending on how they behave; and (2) there is enough resources for all in this world. The main problems are how they are utilized and shared; in spite of the devastation of the resources that are known to us, there are still many that are not known to us or deliberately ignored because of convenience which modern development and progress have become synonymous with. The circular economy can help us deal with all aspects of waste management as well as the exploration of new avenues of survival beyond existing frontiers of our existence, lifestyles, and survival.

5.3.2 Sustainable Finance

Sustainable finance is funding that takes into account all or at least the environment in the three broad ESG categories in making debt or investment decisions. As explained in Section 5.1, it involves expenditure and revenues with an impact on the environment. The vast list of sustainable finance products includes green bonds, green cards, green loans, green mortgages, social bonds, climate bonds, eco deposits, etc.[21]

It aims to create a balance between profitability and economic growth by addressing environmental, social, and/or governance factors in the functioning of institutions, the risks associated with these, and how to address those risks in order to ensure long-term benefits for investors, corporations, organizations, and society at large. For example, the AfDB issued a USD$3.1 billion 3-year 'Fight Covid-19' Social Bond in March 2020. In June 2021, it raised AUS600 million "kangaroo" social bond from the Australian bond market. In June 2022, it launched a 19-billion Ugandan shilling (UGX) ($5.07 million) 2-year theme bond. The proceeds from these bonds are used to alleviate or mitigate social issues through better infrastructure, protection against natural disasters, and creating employment. These are all reflections of the Bank's recent efforts to promote ESG in the African continent through much-needed finance that is sustainable.

[21] See for instance, NATF, Green Financial Products and Services. Current Trends and Future Opportunities in North America. A report of the North American Task Force (NATF) of the United Nations Environment Programme Finance Initiative, Toronto, August 2007.

Commonly, sustainable finance focuses on environmental issues and the financial standards, regulations, and products that they comprise. These environmental issues are climate change adaptation and mitigation as well as the broader issues such as pollution control, biodiversity, water contamination, and circular economy.[22] Increasingly, investors want to understand issuers' exposure to green products and services. This means that if a company has the opportunity to secure or make investments or undertake activities with green components these should be pursued and reflected in the ESG reporting. Such can be beneficial in attracting investors.

Even in Africa, environmentally focused sustainable finance has been around earlier than social bonds. In response to the global trends, the AfDB launched the African Carbon Support Program since November 2010. The aim is to assist African countries to access carbon finance for investment projects.

5.4 Real-Life Examples of ESG Reporting

Next, we provide examples of ESG reports from different countries and institutions and end up with one we have recently designed ourselves for a very rare fund of its kind—the Gambia Special Needs Entrepreneurial Fund. Three of the companies (Perseus, MTN, and A.P. Moller) do all or a significant part of their business in Africa and the authors have experience working with two of the six examples (Perseus and DEWA, the fourth largest Australian gold producer and the biggest company in the Dubai Financial Market, respectively). The most detailed company report that we have found in our research is that of Perseus, a company that the authors work for. As shown in the Table 5.2, "Sustainability Reports" tends to be the more common name for such reports, but "ESG reports" are also used. This is because ESG is a specific enabler of sustainability, particularly as it is known today in the business world, and ESG enables a more organized and comprehensive approach to sustainability. A short answer to the often-asked question about the difference between ESG

[22] https://ec.europa.eu/info/business-economy-euro/banking-and-finance/sustainable-finance/overview-sustainable-finance_en#:~:text=Sustainable%20finance%20refers%20to%20the,sustainable%20economic%20activities%20and%20projects, last visited March 13, 2022.

and sustainability is that ESG is the means and sustainability is the goal or end.

5.4.1 *ESG Reports*

A look at the table clearly shows the illustration of our definition of ESG and sustainability in this book, in Chapters 1 and 3 in particular. We can see from the table the attempts to build bridges between the needs of the present and the future generations and between the environment, society, and governance. In doing so, it is common to align ESG measures to global goals, such as the UN SDGs, most popularly. Additionally, we can see from the table that an ESG/sustainability report can be prepared not only for institutions but even for major projects or events. An example is Expo2020 Dubai, under the theme, "Connecting Minds, Creating the Future", which recorded 24,102,967 visits by attendees from 192 countries in a period of 182 days from 1 October 2021 to 31 March 2022. Bent upon utilizing the multiple benefits of this historic global event, the Dubai Government prepared a 238-page sustainability report addressing ways of assessing and converting the resources, knowledge, and the exhibition centre itself into a long-term experience through ESG.

5.4.2 *The Gambian Special Needs Entrepreneurial Fund (GSNEF): An ESG Framework*

The Gambian Special Needs Entrepreneurial Fund (GSNEF), was established in 2021 from a generous grant from the Arab Bank for the Economic Development of Africa (BADEA) and contribution from Heeno International, a foundation set up in the Gambia and privately funded by the authors, with projects in six African countries. The GSNEF grew out of the Gambia Disability Impact Trust Fund (GDITF) which was conceived in 2017, started in 2017 and initially funded by Heeno for the benefit of the members of the Gambia Federation of the Disabled (GFD). The primary objectives of the projects are to set up a Fund to empower persons with disabilities (or People with Special Needs) in the Gambia with entrepreneurial skills and provide them with seed capital through microcredit in order to undertake viable and sustainable businesses that can reduce their dependency and improve their participation in economic activities. GSNEF provides three types of assistance to the disabled: (1) interest-free microcredit; (2) entrepreneurial training through workshops; and (3) skills training through tailoring apprenticeships. As of 31 December 2022, less than two years after it started operations, GSNEF trained a total of 400 beneficiaries and provided microcredit to others (see Table 5.3). We prepare an ESG reporting framework for the Fund, aligning the ESG issues with the UN SDGs and the UN Global Compact.[23] Based on this framework, annual reports will be prepared to monitor progress. The reporting framework is shown in Table 5.4.

[23] A non-binding agreement under the auspices of the United Nations to encourage businesses to adopt socially responsible practices and disclose progress in their implementation. As of July 2022, 16,540

Companies from 158 countries were members of the pact. For more information visit: https://www.unglobalcompact.org/.

Table 5.4 The ESG reporting framework of GSNEF

	Criteria	Rationale	Metrics	UN SDG	UN Global Compact	Progress	
1	Refill water filter instead of plastic bottles in office	Reduce plastic waste and pollution, which are harmful to the environment	Bottles saved in terms of water consumed	13	Principle 7: Businesses should support a precautionary approach to environmental challenges; Principle 8: undertake initiatives to promote greater environmental responsibility; and Principle 9: encourage the development and diffusion of environmentally friendly technologies.	✗	Want to be transparent
2	Use of second-hand furniture in good shape	Reuse, Repurpose, Recycle Furniture: 'Owning vs Renting'?	Money saved from alternatives	13	Principle 7, 8, 9	✓	
3	Encourage both suppliers and employees to convert to paperless alternatives to be able to cut paper use. Utilise digital platforms.	Reduce paper waste. Environmental Preservation.	Compueter usage, paper saved	13	Principle 7, 8, 9	✗	
4	Encourage suppliers to build the solar panels necessary to increase energy reliance on renewables.	CO2 Emissions reduction. Resource and energy Efficiency.	No of solar installations, reduction in conventional power usage?	13, 9	Principle 7, 8, 9	✗	
5	Work virtually or refurbish an existing building for office headquarters instead of building new one	Reuse, Repurpose & Recycle Principle of Circular Economy. Reducing emissions from land use.	Rent & energy-consumption savings	13	Principle 7, 8, 9	✓	
6	Ensure that suppliers are sustainable when we can	Reducing negative value chain impact.	Supplier Evaluation Scorecard	11, 12, 13	Principle 7, 8, 9	✗	
7	Encourage waste management at home & office	Reduce waste and pollution	Waste management information	13, 12	Principle 7, 8, 9	✗	
8	Hygiene and Health Education	Reduce littering, provide benefits of clean environments, sanitization and have adequate furniture (bed, sofas...).	Hygiene and Health Information	11	Principle 7, 8, 9	✓	
9	Use long life bulbs	Reduce waste. Easier to manage.		12	Principle 7, 8, 9	✗	
						3/9 Complete	

Legend
✗ : Not in progress
✓ : In progress or completed

(continued)

Table 5.4 (continued)

	Criteria	Rationale	KPI	UN SDG	UN Global Compact	Progress
1	Our core mission: create social and financial value by providing training & microcredit to persons of special needs	Economic empowerment of the disabled	No. of people trained, amount of credit disbursed & repaid	10	Principle 1, 2	✓
2	Educational programs about the rights of the disabled	Advocacy and Raising awareness of special needs. Facilitating their integration and acceptance into society.	No. of such progammed & audience	10	Principle 1, 2	✓
3	Media campaign to sensitize and educate the Gambian population about the disabled	Advocacy and Raising awareness of special needs. Facilitating their integration and acceptance into society.	No. of such campaigns & audience reached	16, 10	Principle 1,2	✓
4	Health and Safety working protocols for disabled	To create jobs & better working conditions	No. of employers observing these projects, no. of disabled benefting	3, 8	Principle 4	✓
5	Women inclusion	Equal opportunities. Gender diversity.	Absolute number or % of disabled female employees in companies	1, 10	Principle 1, 2	✓
6	Building social amenities that are sensitive to the needs of persons with special needs	Special Parking, Elevators, Restroom, recreational parks, paygrounds	No. of new facilities & rate of usage	10, 4	Principle 1, 2	✗
7	Educational support for the children of the disabled	Empowering those with special needs	No. of such kids in schools	4	Principle 1, 2	✗
8	Cultural education	Revive cultural awareness for the special needs.	Cultural events organized about values in favour of the disabled and no. of participants	10	Principle 1, 2	✗
9	Equal opportunities for persons with disabilities	To create economic, social and psychological opportunities for the disabled.	No. of persons with disabilities who moved from positions of disadvantaged to advantaged.	10	Principle 1, 2	✓
10	Financial literacy program for the special needs	Financial literacy information and events	No. of trainining events & beneficiaries		Principle 1, 2	✓
						6/10 Complete

	Crieria	Rationale	Metrics	UN SDG	UN Global Compact	Progress
1	Regular business management and skills training for staff	Career development. Capacity Building. Creation of a skilled workforce.		4 QUALITY EDUCATION		3 staff completed an Islamic Finance certification training ✓
2	Appoint independent auditor	Transparent and accurate accouting	Selection process	16 PEACE, JUSTICE AND STRONG INSTITUTIONS		✗
3	Board size and transparency	Independent board, transparency and expertise. Agency Theory (Ross and Mitnick, 1973). Board composition (Lipton and Lorsh,1992; Jensen,1993)	Board composition	16 PEACE, JUSTICE AND STRONG INSTITUTIONS; 17 PARTNERSHIPS FOR THE GOALS; 8 DECENT WORK AND ECONOMIC GROWTH		✓
4	Stakeholder Engagement	Stakeholders are informed & their interests are served. Stakeholder Theory (Freeman, 1984)	Meetings, reports, correspondence with stakeholders	17 PARTNERSHIPS FOR THE GOALS	Principle 6: the elimination of discrimination in respect of employment and occupation	✓
5	Manual Internal Procedures	Rules and regulations to properly run the fund	Existence of such a manual (s)	8 DECENT WORK AND ECONOMIC GROWTH	Principle 6: the elimination of discrimination in respect of employment and occupation	✓
6	Good relations with donors	Making sure to keep them informed about the project status	Meetings, reports, correspondence with donors	17		✓
7	Create Website	Keeping stakeholders and potential investors informed	Existence of a website	17		✗
8	Anti-corruption	Ensure the absence of corruption	Anti-corruption disclosure (0,1)?	16 PEACE, JUSTICE AND STRONG INSTITUTIONS; 17 PARTNERSHIPS FOR THE GOALS	Principle 10: Businesses should work against corruption in all its forms, including extortion and	✓
						6/8 Complete

PART III

Regional Case Studies

CHAPTER 6

ESG in Egypt, Kenya, Nigeria, and South Africa

Abstract This chapter carries out a survey of the reality and potentiality of ESG in Africa's main regions—East, West, South, and North. The biggest economy in each region is looked at through its international agreements, national laws and stock markets, and NGOs. The interest and activity in each of the countries are found to be high and increasing, although recent, and affected by global developments.

Keywords Stock markets · NGOs · Egypt · Kenya · Nigeria · South Africa

6.1 Introduction

This chapter covers four of the biggest economies in Africa, one from each subregion of the continent. They are: Egypt (North Africa), Kenya (East Africa), Nigeria (West Africa), and South Africa (Southern Africa). Starting with an introduction to each economy, we address the following questions in each of the countries studied:

1. What are the national laws which directly support or can support ESG reporting?

K. N. M. Sonko and M. Sonko, *Demystifying Environmental, Social and Governance (ESG)*, Palgrave Studies in Impact Finance,
https://doi.org/10.1007/978-3-031-35867-8_6

2. What are the international agreements/treaties which the government is party to, which directly support or can support ESG reporting? In addressing this question, we delve into relevant international agreements that each country has signed in each category of ESG from the late nineteenth century to date.
3. Which are the biggest locally owned and foreign companies in the country and do they already subscribe to ESG reporting?[1]
4. Is there a stock exchange which provides for ESG reporting?
5. Are there active non-governmental organizations (NGOs) which directly support or can support ESG reporting?

The aim of the chapter is to determine whether or not the country already has the potential legal/regulatory framework at the national and international levels and the domestic institutions (public and private) which have already laid or can lay down the foundations for ESG reporting in the country. With regard to the laws and treaties, the aim is to also explore whether the country can justify setting up comprehensive ESG reporting nationally and join the rest of the world in this important initiative, based on existing local laws and international agreements. A further purpose of this exercise is to imply that ESG should not be limited to the private sector alone but be extended to governments themselves. In what follows, it is extensively shown that African governments, as illustrated by the case studies in this chapter, already have extensive and well-established domestic and international structures, covering all aspects of ESG reporting, on earth including activities on the ground, in the air and oceans, and extending very far into space, encompassing all three ESG categories. In a way, the governments have indirectly cut the work of the companies for them as the lists can indicate priority areas for ESG actions at the national levels too. As Refaat (2014) pointed out in her study of CSR in Egypt, annual CSR and Sustainability reports may also present the companies' efforts towards the environment, surrounding community, and its responsible business practices, including respect for existing laws and rights in general.

The Egyptian Investment Law goes further by stating that the voluntary actions of companies should respect national and international

[1] Due to data constraints, we rely mainly on stock market figures and hence listed companies.

"conventions, executive regulations, declarations, resolutions, guidelines and charters".[2] Our very long lists of agreements/treaties cover almost all the relevant ones dating back to the colonial period in Africa.

The discussions below are in standard format for each country, including South Africa which is covered more elaborately in the empirical study in the next chapter. This standard format provides structure and consistency in the chapter and information provided.

6.2 Egypt

6.2.1 A Glimpse of the Economy

According to World Bank data, the Egyptian economy grew from $20.3 billion in 1960 to $425.9 billion in 2021, based on GDP in constant 2015 US dollars.[3] However, GDP performance during the period was inconsistent, from a growth rate of 5.2% in 1961 to negative 1.6 in 1968, up by 13.3% in 1976, up by 1.1% in 1991, 1.8% in 2011, and up again by 3.3% in 2021, for examples.[4] The unemployment rate stood at 9.3% in 2021.[5] Unlike most African countries, the Egyptian economy depends on industry, with manufacturing accounting for the biggest share of the economy. The biggest contributors to GDP in 2021 are:

1. Manufacturing industries 16.1%
2. Wholesale and retail trade 13.6%
3. Agriculture, forestry, and fishery 11.3%
4. Real estate activities 10.3%
5. Mining 9.7%[6]

[2] Fady Tawakol, "Corporate Social Responsibility (CSR). Its Contribution in Achieving Sustainable Development in Egypt 2030 in Light of the Unified Investment Law No. 72 of 2017," https://espesl.journals.ekb.eg/article_213580_a2503bf7ca8475a28a779d7d2b9a727a.pdf.

[3] https://data.worldbank.org/indicator/NY.GDP.MKTP.KD?locations=EG-KE-NG-ZA. All dollar figures throughout this book refer to United States dollars unless otherwise stated.

[4] https://data.worldbank.org/indicator/NY.GDP.MKTP.KD.ZG?locations=EG-KE-NG-ZA.

[5] Source for the unemployment rate: Egypt—Unemployment rate 2021 | Statista https://www.statista.com.

[6] See https://www.oecd-ilibrary.org/sites/0628d72c-en/index.html?itemId=/content/component/0628d72c-en, last visited November 19, 2022.

Egypt's social, economic, and political situations in the last decade, particularly since the downfall of the late President Mubarak in 2011, posed obstacles for local businesses to expand. However, even before 2011, Egypt's economy had taken a downturn, especially after the 2008 Global Financial Crisis, notably characterized by a hike in food prices. It took a harder hit after the 2011 revolution, when tourism, an important sector of the Egyptian economy, plummeted because of political crisis and security concerns. As a consequence, foreign exchange reserves fell from $36 billion in December 2010 to only $16.3 billion in January 2012. The government had to resort to a $12 billion, three-year International Monetary Fund loan in 2016, as part of efforts to revive the economy. Since then, Egypt has been trying to woo foreign investors.

6.2.2 Relevant National Laws and ESG Reporting

6.2.2.1 Environmental

Environmental issues in Egypt are governed by Law No. 4 of 1994. This law provides for the creation of an agency (the Egyptian Environment Affairs Agency [EEAA]) for the protection and promotion of the environment. The EEAA formulates general policy and prepares the necessary plans for the protection and promotion of the environment. It also follows up on the implementation of such plans. The law provides for mandatory environmental reviews to be undertaken by the competent administrative/governmental authority according to EEAA's instructions, as part of the approval process for all new projects.

The law forbids the handling of hazardous materials and wastes or the construction of any establishment for the treatment of such substances without a license from the competent administrative/governmental authority. It is prohibited to incinerate, to dispose of, or to treat garbage and solid wastes, as well as to spray pesticides or any other chemical compound, unless it is done according to the conditions and safety measures specified in the Executive Regulations of the law. It is also forbidden to import hazardous waste or to allow its entrance into or passage through Egyptian territories. It is mandatory for all those who produce or handle dangerous materials to take precautions to ensure that no environmental damage shall occur. All establishments (industrial and others) are required to ensure that whilst undertaking their activities, no leaked or emitted air pollutants (caused by the burning of fuels, etc.) shall exceed the maximum permissible levels.

6.2.2.2 *Corporate Social Responsibility*

There are no mandatory laws or state regulations for CSR in Egypt.[7] On the contrary, the government uses incentives to encourage companies, especially international ones investing in Egypt, to engage in corporate social responsibility projects.[8] The 2016 Corporate Governance Code encourages companies to clearly set their policies on economic, social, and environmental responsibilities. This should include initiatives, objectives, scopes, and benefits to stakeholders.

Article 15 of the Unified Investment Law, No. 72, 2017, also recommends that in order to contribute "Toward achieving the goals of comprehensive and sustainable development, the Investor may dedicate a percentage of his annual profits to create a social development system, outside of his Investment Project", which would be used for the following purposes: (1) protection of the environment; (2) providing services or programmes in the areas of healthcare, social care, or cultural care, or other development areas; (3) supporting technical education or the funding of research, studies, and the awareness campaigns aiming at developing and improving the production, in agreement with any of the universities or scientific research institutions; and (4) training and scientific research. The law prohibits the use of CSR projects for political, religious, and discriminatory purposes.[9]

6.2.2.3 *Corporate Governance*

Corporate governance has largely been based on guidelines issued by the Egyptian Institute of Directors (EIoD). In 2005, the Institute introduced a Corporate Governance Code for listed companies to help them comply with disclosure requirements for their shareholders and pursue international levels of sustainability. In 2006, the EIoD came up with

[7] Refaat (ibid.), https://core.ac.uk/download/pdf/333724443.pdf, last visited December 24, 2022.

[8] See, for instance, the Ministry of investment and international cooperation's contests leading companies in the field of corporate social responsibility, https://www.investinegypt.gov.eg/English/NewsAndEvents/News/Pages/Corporate-social-responsibility.aspx, last visited December 24, 2022.

[9] Law No. 72 of 2017, Promulgating the Investment Law, https://www.gafi.gov.eg/english/startabusiness/laws-and-regulations/publishingimages/pages/businesslaws/investment%20law%20english%20ban.pdf, last visited December 24, 2022.

another Corporate Governance Code, this time for State-Owned Enterprises (SOEs), affirming the role of the State and the controls and supervision needed for the effective functioning of these enterprises. In 2011, the EIoD revised the Corporate Governance Code for listed companies in order to bring it at par with regional and international best practices, especially with regard to prudent and transparent management for the benefit of all stakeholders, and incorporate the roles of the Board of Directors and the subject of sustainability. The principle of "Comply or Explain" was also introduced. The updated version of the 2011 Code as approved and issued by the Egyptian Financial Supervisory Authority (EFSA)'s Board of Directors Resolution No. 84 of 26 July 2016.

6.2.3 *Relevant International Agreements/Treaties and ESG Reporting*

This subsection presents the outcome of our research on international agreements which Egypt might have already signed in each of the three categories of ESG. As discussed earlier, the objectives are to show whether the country already has international agreements (in addition to the national laws discussed above) for justifying and setting up comprehensive ESG reporting nationally, responding to the international calls for ESG reporting and using these international agreements to reinforce the national cases for ESG disclosures. It is shown in all cases and ESG categories that the potential legal/regulatory frameworks and justifications exist and have been there for a very long time, although they were not in relation with, or intended for, comprehensive ESG as it is known today. For a detailed and comprehensive list of the agreements/treaties, the reader is referred to Appendix Table 6.9.

6.2.4 *The Biggest Locally Owned and Foreign Companies*

We ask in this section (as in the case of the other cases in this chapter) "which are the biggest locally-owned and foreign companies in Egypt by market capitalization, turnover or any other economic indicator and do they already subscribe to ESG disclosures or reporting?" In addition to the macro/country-level surveys in this chapter, we now provide information at the micro/company level in Egypt. The same is done in each of the other three countries that follows.

Based on market values, sales, assets, and profits, Forbes identifies the following companies as the top listed companies in Egypt (Table 6.1).

6.2.5 The Egyptian Stock Exchange

Egypt has a stock exchange—the Egyptian Stock Exchange (EGX)—established since 1883, making it the oldest stock exchange in Africa. It is the fourth biggest in Africa, with market capitalization averaging $41.35 billion from 2006–2020.[10] As one of the four founders of the United Nations' Sustainable Stock Exchanges initiative (SSE) in 2009, the Exchange is an early promoter of ESG reporting. In 2010, the Environment, Social and Governance Index for Egypt (The S&P/EGX ESG Index) was created by Standard & Poor (S&P) Dow Jones Indices and the Egyptian Stock Exchange. This is the first and only ESG index in the Middle East and North Africa. The Egyptian Exchange does the ESG score research and management, whilst S&P Dow Jones Indices provides the methodology for defining relevant ESG criteria and does the calculation for the index.

The purpose of the S&P/EGX ESG index is to raise the profile of those companies that meet the Eligibility Criteria and perform the best through scores in the three categories of ESG in comparison with their market peers. There are 30 of such companies on the Egyptian Exchange. They were selected in two steps: the first was by creating an ESG score for each of a 100 companies in the Index Universe; and the second by creating the index of companies with the highest scores. The weight of each company in the second index depends on a variety of factors, such as its relative score, size, liquidity, etc.[11]

EGX is a member of the Sustainability Working Group (SWG), set up in 2014 by the World Federation of Exchanges. From 2016–18, EGX was the Vice Chairman of the Group. The Exchange has set up an Internal Committee to manage its own CSR activities for raising and donating charitable funds from the EGX, employees and companies, in cooperation with the Egyptian Ministry of Social Solidarity. The Exchange publishes announcements that it is particularly interested in, and will, facilitate

[10] https://www.theglobaleconomy.com/rankings/stock_market_capitalization_dollars/, last visited October 31, 2022.

[11] For more details refer to S&P Dow Jones Indices, *S&P/EGX ESG Index Methodology*, January 2021.

Table 6.1 The biggest companies in Egypt

Name	*Market value (US$, millions)*	*ESG/sustainability reporting*	*Origin*
1. Commercial International Bank (CIB)	4.5 billion	Yes https://enterprise.press/wp-content/uploads/2020/11/CIB-5-Years-of-Sustainability-Reporting-PR-E-V5-Legal-converted.pdf	Joint venture between the National Bank of Egypt and the Chase Manhattan Bank
2. QNB ALAHLI	2 billion	Yes https://www.qnbalahli.com/sites/qnb/qnbegypt/page/en/enesgreportsanddisclosures.html	Qatari QNB Group acquired the majority stake of Egypt's NSGB Bank and changed its name to QNB ALAHLI
3. Telecom Egypt	1.5 billion	Yes https://www.te.eg/wps/portal/te/About/CSR/	Egypt
4. Elsewedy Electric	878 million	Yes https://ir.elsewedyelectric.com/en/sustainability-overview	Elsewedy family
5. Talaat Moustafa Group Holding (TMG Holding)	953 million	Yes https://www.csrhub.com/CSR_and_sustainability_information/Talaat-Moustafa-Group-Holding	Egypt
6. Al Ezz Dekheila Steel Co.—Alexandria (EZDK)	562 million	Yes https://www.ezzsteel.com/about-ezz-steel/sustainability	Egypt
7. Orascom Construction	464 million	Yes https://orascom.com/sustainability/	Established more than 72 years ago by late billionaire Onsi Sawiris
8. Ezz Steel	408 million	Yes https://www.ezzsteel.com/about-ezz-steel/sustainability	Egypt
9. Eastern Company	1.3 billion	Yes https://eastern.in/csr-policy/	Egypt
10. Faisal Islamic Bank of Egypt (FIBE)	532 million		Egypt

(continued)

Table 6.1 (continued)

Name	*Market value (US$, millions)*	*ESG/sustainability reporting*	*Origin*
11. Emaar Misr	658 million	Yes https://www.emaarmisr.com/en/about-emaar/emaar-csr/ https://cdn.properties.emaar.com/wp-content/uploads/2021/07/EMAAR-PROPERTIES-ESG-2020.pdf	UAE
12. Egypt Kuwait Holding (EKH)	1.4 billion	Corporate Citizenship https://www.ekholding.com/Corporate-Citizenship.aspx	Egypt-Kuwait
13. Misr Fertilizers Production Company (MOPCO)	1.2 billion	Yes http://mopco-eg.azurewebsites.net/sustainability	Egypt
14. EFG Hermes Holding	860 million	Yes https://efghermes.com/en/sustainability-strategy	Egypt
15. Abu Qir Fertilizers and Chemicals Company (AFC)	1.7 billion	Yes http://abuqir.net/uploads/sustainability/AbuQir_Sustainability_2020.pdf	Egypt
16. Housing & Development Bank (HDB)	303 million	Yes https://hbtf.com/en/news/sustainability-report	Egypt
17. GB Auto Group	228 million	Yes https://ir.ghabbourauto.com/en/sustainability-reports	Egypt
18. Crédit Agricole Egypt	442 million	Yes https://www.ca-egypt.com/en/corporate-social-responsibility/ca-egypt-foundation-development/	France
19. Abu Dhabi Islamic Bank—Egypt (ADIB)	136 million	Yes https://adib.ae/en/SiteAssets/Investor-Relations/ESG-reports/ADIB-ESG-Report-2021_EN.pdf https://www.adib.eg/about-adib-egypt/csr	UAE

(continued)

Table 6.1 (continued)

Name	*Market value (US$, millions)*	*ESG/sustainability reporting*	*Origin*
20. Oriental Weavers Carpet	269 million	Yes https://orientalweavers.com/sustainability/	Egypt

As in subsequent tables of the biggest companies and of NGOs, the companies here are listed neither alphabetically nor in terms of size. The reader should further note that market values do change
Additional Sources Forbes, "Top 50 Listed Companies in Egypt 2022," https://www.forbesmiddleeast.com/lists/top-50-listed-companies-in-egypt-2022/, last visited 19 November 2022; and various

the listing process for companies in eco-industries—renewable energy, waste recycling, transportation, etc., through IPOs, bonds, Sukuk, and infrastructure funds.

In 2021, the Egyptian Financial Regulatory Authority (FRA), in an effort to put businesses in line with global trends and initiatives that seek to minimize climate risks, issued resolutions requiring all companies listed on the Stock Exchange and companies operating in non-bank financial activities to submit ESG disclosure reports. The FRA has completed the setting up of the regulatory framework for companies preparing disclosure reports on their Environmental, Social, and Governance practices. This came after it succeeded in completing the approval of the first issuance of green bonds (July 2021) in the Egyptian capital market, with a value of $100 million for one of the listed companies. Resolutions Nos. 107 and 108 (2021) require listed companies and companies operating in non-bank financial activities to submit Environmental, Social, and Governance disclosure reports. This is intended to make the ones with issued capital or net ownership rights of not less than 500 million Egyptian pounds[12] to complete disclosures related to the financial effects of climate change

[12] 1 USD = 24.6 EGP on December 18, 2022, https://www.exchangerates.org.uk/Dollars-to-Egyptian-Pounds-currency-conversion-page.html, last visited December 18, 2022.

in line with the Task Force on Climate-Related Financial Disclosures (TCFD).[13]

6.2.6 *Influential Non-Governmental Organizations (NGOs)*

Non-governmental organizations and charities in Egypt, like those all over the world, rely on the generous support of the public to be able to carry out their work. As in other Muslim communities and Islamic countries, the holy month of Ramadan is a time when Muslims typically give alms for the poor and needy in the form of charity (*sadaqah*) and Islamic tax in the form of *zakat*. These practices and Ramadan can reveal the most active NGOs that are involved in economic, social, and religious activities. *Al-Masry Al-Youm* and the *Egypt Independent* newspapers looked at the list of charities and institutions that the Egyptian public have supported the most in recent years during the month of Ramadan in discharging their charitable duties and they came up with the "Top Five" most influential Egyptian NGOs below.[14] Other active NGOs are shown in Table 6.2.

1. Egypt Foundation for Youth & Development, https://www.egyptfoundation.org/
2. Together Association for Development and Environment, https://www.together-eg.com/
3. The Egyptian Food Bank, https://www.globalgiving.org/projects/egyptian-food-bank/
4. Egyptian Association for Informal Development and Environment Protection,
5. 57357 Children's Cancer Hospital Foundation, https://www.57357.org/en/home-page/.

[13] For more information refer to SSE, "Egyptian FRA: Mandatory ESG and Climate Disclosure Regulation," 15 July 2021, https://sseinitiative.org/all-news/egyptian-fra-issued-mandatory-esg-and-climate-disclosure/, last visited December 8, 2022; and FRA, "Companies listed on the Egyptian Exchange and engaged in non-banking financial activities request new disclosure reports on sustainability practices and the financial impacts of climate change," July 13, 2021, last visited December 8, 2022.

[14] See https://egyptindependent.com/5-most-popular-charities-donated-egypt-2014-2016/, last visited December 8, 2022.

Table 6.2 Selected list of NGOs in Egypt

	Name	*Website*
1	Abdel Hady Association for Charity	http://www.nccm-egypt.org/e3/e1591/e2693/index_eng.html
2	Al Kasr Al Ainy Friends Association—KEFA	https://yellowpages.com.eg/en/profile/afnci-_-association-of-friends-of-the-national-cancer_free-iniative/395885
3	Arab Association for Human and Environmental Development	https://yellowpages.com.eg/en/profile/arab-association-for-human-&-environmental-development/305282
4	Leaders Egyptian Association For Development "LEAD"	https://www.annalindhfoundation.org/members/leaders-egyptian-association-development-lead
5	Child Neurology Association	https://www.childneurologysociety.org/
6	Education Development Association	https://www.eda.edu.au/
7	Egyptian Red Crescent Association	https://www.egyptianrc.org/
8	EL Mabarra Association in Alexandria	https://www.cybo.com/EG-biz/el-mabarra-assoc-semouha-nursery
9	Health and Environment Development Association—AHCD	https://yellowpages.com.eg/en/profile/association-of-health-&-environmental-development/178783
10	International Youth Foundation	https://iyfglobal.org/country/egypt
11	Institute of Cultural Affairs—ICA	https://www.ica-international.org/ica-worldwide/ica-mena-egypt/
12	Manshiyat Nasser Pioneers Association The National Council for Childhood and Motherhood	http://www.nccm-egypt.org/e3/e1591/e2693/index_eng.html
13	New Horizon Association for Social Development (NHASD)	https://www.facebook.com/New-Horizon-Association-for-Social-Development-NHASD-199004996789775/
14	Social and Cultural Clubs Association	http://www.escd-egypt.org/en/DynamicPages.aspx?page_id=2021
15	Union of Prevention from Addiction	https://www.narconon-egypt.org/en/

Companies can cooperate with NGOs in Egypt and other African countries, to execute their ESG programmes. We would like to stress that locally owned NGOs should be given preference for such partnerships, as this can help in local capacity building in the NGOs and help companies, especially the foreign ones, identify local needs better through such NGOs. However, in areas where actual or potential local NGO capacity does not exist, cooperation should be with international NGOs, which are included in the lists in this chapter. In all cases, companies should look at the synergy between the activities and objectives of the chosen NGO and their own. This should be the case in Egypt and throughout Africa.

6.3 Kenya

6.3.1 A Glimpse of the Economy

According to World Bank data, the Kenyan economy grew from $6.1 billion in 1960 to $90.37 billion in 2021, based on GDP in constant 2015 US dollars.[15] However, GDP performance during the period has been erratic, from a negative growth rate of 7.8% in 1961, up by 22.2% in 1971, dropped by a negative 0.8% in 1992, up by 3.8% in 2013, and up again by 7.5% in 2021, for examples.[16] The unemployment rate stands at 5.73% in 2021.[17] Unlike most African countries, the Kenyan economy depends on the services sector, not agriculture. Kenya's financial sector is now the third largest in Sub-Saharan Africa, after South Africa and Nigeria. In Kenya the biggest contributing sector to GDP in 2021 are:

1. Services 53.9%
2. Agriculture, forestry, and fishing 22.6%
3. Industry 17.4%
4. Manufacturing 7.6%.[18]

Although Kenya's economic outlook is generally one of the most positively viewed in Africa, it is subject to most of the same challenges faced by other countries throughout Africa. In the midst of the COVID-19-afflicted years (when the pandemic was at its height in particular),

[15] https://data.worldbank.org/indicator/NY.GDP.MKTP.KD?locations=EG-KE-NG-ZA. All dollar figures throughout this book refer to United States dollars unless otherwise stated.

[16] https://data.worldbank.org/indicator/NY.GDP.MKTP.KD.ZG?locations=EG-KE-NG-ZA.

[17] Source for the unemployment rate: Egypt—Unemployment rate 2021 | Statista https://www.statista.com.

[18] World Bank, World Development Indicators: Structure of value added: http://wdi.worldbank.org/table/4.2, last visited November 1, 2022.

these challenges include greater economic uncertainty, exposure (as a net wheat, fertilizer, and fuel importer) to the unfavourable price impacts of the Russia-Ukraine war, difficulties in maintaining productive capacities, and inadequate containment measures for the pandemic. A new government that was voted into office in August 2022 now aims to align Kenya's long-term development goals to Vision 2030 in order to transform Kenya into a competitive, prosperous, and people-oriented country, through a bottom-up or inclusive economic growth model emphasizing manufacturing, agriculture, housing, and healthcare.[19]

6.3.2 Relevant National Laws and ESG Reporting

6.3.2.1 Environmental

The Kenyan government has taken some steps to protect the environment over the years. The biggest efforts have been in the area of halting environmental degradation. It has committed to the planting of 1.8 billion trees by 2022.[20] The National Environmental Management Authority deals with environmental matters across the country. It has the mandate to monitor, assess, and audit the impact of different activities on the Kenyan environment.

The 2010 constitution of Kenya was hailed as a progressive one for dealing with issues rarely addressed by Kenyan national law before, such as difficult or sensitive ones relating to land tenure and environmental degradation. It states that the state has an obligation to protect the environment as well as protect the rights of the people in their dependence on and usage of the environment. Article 69 of the new constitution outlines the obligations of the state in protecting the environment, which include, but are not limited to, the following:

i. Ensuring sustainable utilization, management, and conservation of the environment and natural resources, and ensuring the equitable sharing of the accruing benefits;

[19] The World Bank, "The World Bank in Kenya," October 3, 2022: https://www.worldbank.org/en/country/kenya/overview.

[20] Boris Ngounou, "KENYA: Government Reaffirms Vision of Planting 1.8 Billion Trees by 2022," Afrik 21, published on May 25, 2020/Modified on May 25, 2020.

ii. Achieving 30% forest cover of Kenyan landmass, after the achievement already of the constitutional 10%[21];
iii. Protecting the indigenous communities' knowledge in biodiversity and genetic resources;
iv. Protecting genetic resources and biodiversity;
v. Prohibiting activities that are harmful to the environment;
vi. Establishing systems to monitor, audit, and assess impacts on the environment by various activities;
vii. Ensuring that gains from the environment benefit all Kenyans.

Article 42 of the constitution guarantees a clean and healthy environment for the people. This includes the right to have the environment protected in the present and future generations. Article 70 expands on this by stating that anyone who feels these rights have been denied, infringed, violated, or threatened, in any way, can seek legal redress in a court of law. The court as a state agency has an obligation to stop, prevent, and discontinue any act of commission or omission that is harmful to the environment. It has the powers to compel a state officer to take action to stop harmful environmental actions by anyone. Where the case can be proven that the victim's rights were indeed violated or denied, the court can order compensation or any redress it sees fit.

The parliament also passed the Mining Act (2016). This act guides the exploitation of mineral resources and the sharing of the benefits. The act states that minerals are part of the land held in trust by the government and as such the government has the obligation to ensure equitable distribution of the gains from these resources, and further protect the land during their exploitation for the benefit of Kenyans.

6.3.2.2 Corporate Social Responsibility

In Kenya, as in the other countries in this study and in Africa generally, CSR is considered as an important but voluntary gesture by companies. However, changing international trends are inserting pressures to do

[21] UNDP, "President Uhuru sets an Ambitious 30% Target for Forest Cover by 2050 during the Launch of Kenya's Tree Growing Fund and Campaign," May 31, 2022, https://www.undp.org/kenya/press-releases/president-uhuru-sets-ambitious-30-target-forest-cover-2050-during-launch-kenya%E2%80%99s-tree-growing-fund-and-campaign, last visited December 18, 2022.

more, as Dixon (2019) observed: "CSR in Kenya has long been characterised by voluntary actions of a philanthropic nature, particularly directed towards challenges of persistent poverty, poor access to education, health care, water and sanitation facilities, as well as food insecurity. However, there has been a slow shift to addressing employee and environmental issues, in the pursuit of sustainability, as reflected in global trends."[22]

6.3.2.3 *Corporate Governance*

In 2002, the Capital Markets Authority (CMA) issued the "Guidelines on Corporate Governance Practices by Publicly Listed Companies in Kenya". In 2016, the CMA replaced it with a code for corporate governance practices for publicly listed companies. As the cases of the other African countries in this chapter show, the 2016 Code was in response to the changing international business environment coupled with the desire to align local standards with international best practice in order to enhance institutional strengthening for listed companies. The Code outlines principles and recommendations in relation to processes and structures necessary for corporate culture and governance. It calls for higher standards than the minimum legal requirements, based on an "Apply or Explain" principle. It obliges boards to report and explain any non-compliance at their Annual General Meetings and in their annual reports. The Code sets out details in relation to executive pay provisions; age limit for board members; directors' terms; board diversity; shareholders; and stewardship.[23]

6.3.3 *International Agreements/Treaties and ESG Reporting*

As in the case of Egypt, this subsection shows the outcome of our research on international agreements which Kenya might have already signed in each of the three ESG categories. The objectives are to show whether the country already has international agreements (in addition to the

[22] Lorraine Dixon, "The Evolution of Corporate Sustainability in Kenya," Kenya Organization for Environmental Education (KOEE), September 4, 2019, https://koeeorg.wordpress.com/2019/09/04/the-evolution-of-corporate-sustainability-in-kenya/, last visited December 24, 2022.

[23] For more information, the interested reader may refer to ECGI, "Corporate Governance in Kenya," https://ecgi.global/content/corporate-governance-kenya#:~:text=The%20code%20sets%20out%20the,the%20minimum%20prescribed%20by%20legislation, last visited December 28, 2022.

national laws discussed above) for justifying and setting up comprehensive ESG reporting nationally, responding to the international calls for ESG reporting, and using these international agreements to reinforce the national cases for ESG disclosures. It is shown in all cases and ESG categories that the potential legal/regulatory frameworks and justifications in Kenya exist and have been there for a very long time, although they were not in relation with, or intended for, comprehensive ESG as it is known today. For a detailed and comprehensive list of the treaties, the reader is referred to Appendix Table 6.10.

6.3.4 The Biggest Locally Owned and Foreign Companies

As the Kenyan economy grew over the years, making it one of the largest economies in Africa, according to its GDP, companies worldwide have chosen to make the country their base so that they can have access to the abundant opportunities in the robust national and regional markets. The wealthiest locally owned and foreign businesses operating in Kenya include those shown in Table 6.3.[24]

6.3.5 The Nairobi Stock Exchange and ESG Reporting

The Nairobi Securities Exchange (NSE) ranks the fifth largest exchange in Africa with a market capitalization of $17.19 billion in 2022.[25] Since joining the Sustainable Stock Exchanges (SSE) initiative in 2015, the NSE has started publishing statements about ESG and sustainability. In 2016, it announced plans to launch Kenya's first sustainability index to encourage company-investor dialogue and listed firms to adopt better ESG practices. Nevertheless, progress appears to be slow, with ESG reporting not yet required as a listing rule, no written guidance on ESG reporting and no sustainability-related index so far. Kenya's Capital Markets Authority has raced ahead of its regional peers with its Stewardship and Corporate Governance Codes, in what could guide the

[24] https://www.statista.com/statistics/1314036/leading-companies-in-kenya-by-revenue/, https://www.nse.co.ke/listed-companies/, https://marcopolis.net/kenya-top-10-companies/ and https://kenyanmagazine.co.ke/richest-companies-in-kenya/.

[25] AskTraders, "5 Biggest Stock Exchanges in Africa," https://www.asktraders.com/za/learn-to-trade/stock-trading/biggest-stock-exchanges-in-africa/Updated, August 15, 2022.

Table 6.3 The biggest companies in Kenya

Name	*Market cap (US$, millions)*[26]	*ESG/sustainability reporting*	*Origin*
1. Safaricom	15 billion	Yes https://www.safaricom.co.ke/sustainabilityreport_2020/	Kenya
2. Equity Group Holdings	1.99 billion	Yes https://equitygroupholdings.com/wp-content/uploads/2019/11/environmentalandsocialriskmanagementpolicy.pdf	Kenya
3. Kenya Commercial Bank KCB Group	1.49 billion	Yes https://bi.kcbgroup.com/fr/br-contact-us/2-general/147-kcb-sustainability-management	Kenya
4. East African breweries	1.29 billion	Yes https://www.eabl.com/sites/default/files/inline-files/EABL_Sustainability_Report_2021_1.pdf	Kenya
5. Co-operative Bank of Kenya	751 million	Yes https://www.co-opbank.co.ke/investor-relations/sustainability-reports/	Kenya
6. Absa bank Kenya	458.907 million	Yes https://www.absabank.co.ke/sustainability/	Kenya
7. Standard chartered bank Kenya	566 million	Yes https://www.sc.com/en/sustainability/resource-hub/	UK

[26] https://www.exchangerates.org.uk/Dollars-to-Kenyan-Shillings-currency-conversion-page.html#:~:text=Dollar%20to%20Kenyan%20Shilling%20Exchange,Convert%20Dollars%20to%20Kenyan%20Shillings. As of April 2023: 1 USD = 134.8995 KES.

Name	*Market cap (US$, millions)*	*ESG/sustainability reporting*	*Origin*
8. British American tobacco Kenya	442 million	Yes https://www.bat.com/sustainabilityreport	US-UK
9. NCBA Group	427 million	Yes CSR https://ncbagroup.com/wp-content/uploads/2021/07/NCBA-GROUP-CORPORATE-GOVERNANCE-REPORTING-TEMPLATE-2021.pdf https://www.ncbal.com/corporate-social-responsibility.php	Kenya
10. I&M Holdings	354 million	Integrated statement https://www.imbankgroup.com/wp-content/uploads/2022/05/IM-Group-Plc-Annual-Intergrated-Report-Financial-Statements-2021.pdf	Kenya
11. Nation Media Group	24.759 million	Yes https://www.nationmedia.com/sustainability	Kenya
12. Carbacid Investments	24.907 million	Yes https://carbacid.com/documents/47/CIL_Sustainability_Report-_FInal_October_2022.pdf	Kenya
13. Bamburi Cement	89.844 million	Yes https://www.lafarge.co.ke/sustainability	Kenya
14. Bank of Kigali BK Group Kenya	197.183 million	Investor Relations https://bk.rw/investor-relation	Rwanda
15. Stanbic Holdings SBIC	279.614 million	Yes https://reporting.stanbicibtc.com/about-us/sustainability	
16. Kenya Power (KPLC)	22.016 million	Evidence unavailable https://kplc.co.ke/content/item/1484/corporate-profile	Kenya
17. Kenya Airways Limited	180.652 million	Yes https://www.kenya-airways.com/.../Sustainability_Policy_updated.pdf	Kenya

(continued)

Table 6.3 (continued)

Name	*Market cap (US$, millions)*	*ESG/sustainability reporting*	*Origin*
18. Old Mutual Kenya	23.424 million	Yes https://www.marketscreener.com/quote/stock/OLD-MUTUAL-LIMITED...	
19. Centum Investment Company	38.991 million	Integrated Report https://centum.co.ke	Kenya
20. Total Energies Marketing Kenya	32.468 million	Yes https://totalenergies.com/sustainability	Kenya

Sources https://victormochere.com/top-10-most-valuable-companies-in-kenya; https://markets.ft.com

development of a regional code in East Africa.[27] Accompanied by the 2015 Companies Act, which introduced a new corporate legal framework, the Codes promote investor-corporate engagement, and suggest that investors should encourage companies to operate with positive social impacts. Investors should also incorporate social, environmental, and ethical concerns into their investment decisions.

6.3.6 *Influential Non-Governmental Organizations (NGOs)*

Kenya has over 10,000 NGOs, some of which have not fulfiled all registration requirements.[28] The long list includes international ones from foreign countries such as the United Kingdom and Canada. The majority of them focus on the environment, health, and human rights; in other words issues directly relevant to ESG. The main NGOs are shown in Table 6.4.[29]

6.4 Nigeria

6.4.1 *A Glimpse of the Economy*

According to World Bank data, the Nigerian economy grew from $64.72 billion in 1960 to $511.93 billion in 2021, based on GDP in constant 2015 US dollars.[30] Overall GDP growth rate on a year-on-year basis during the period has not been very impressive, from a negative growth rate of 0.2% in 1961, down by negative 5.2% in 1975, down by a negative 10.83% in 1983, up by 15.3% in 2002, up by 5.6% in 2006, again

[27] World Bank, "The World Bank in Eastern and Southern Africa," https://www.worldbank.org/en/region/afr/eastern-and-southern-africa#:~:text=South%20Africa%2C%20an%20upper%20middle,by%20Angola%2C%20Kenya%20and%20Ethiopia, last updated: October 29, 2020.

[28] TUKO, "Top 15 NGOs in Kenya 2022 and Their Contacts: Best Registered NGOs," https://www.tuko.co.ke/338191-list-ngos-kenya-contacts.html, last visited December 12, 2022.

[29] *Sources* See https://www.tuko.co.ke/405443-list-hospitals-kiambu-county-private-public.html, last visited January 5, 2023; and https://projectworldimpact.com/country/Kenya?gclid=EAIaIQobChMIwqSp0oux_AIVCbrtCh08lQ3eEAAYBCAAEgLrAvD_BwE, last visited January 5, 2023.

[30] https://data.worldbank.org/indicator/NY.GDP.MKTP.KD?locations=EG-KE-NG-ZA. All dollar figures throughout this book refer to United States dollars unless otherwise stated.

Table 6.4 Selected list of NGOs: Kenya

	Name	*Website*
1	African Development and Emergency Organization	www.adeo.or.ke
2	Africa Harvest Biotech Foundation International	https://africaharvest.org/
3	Aga Khan Foundation (AKF)	https://the.akdn/en/how-we-work/our-agencies/aga-khan-foundation
4	Amref Health Africa	https://amref.org/
5	Center for Enterprise Development & Innovation (CEDI)	https://cedinnov.org/
6	Clean Start Kenya	https://cleanstartkenya.org/
7	Crisis Pregnancy Ministries	https://hopepregnancyministries.org/
8	East Africa Partnership	https://www.eastafricapartnership.org/
9	Engender Health Ltd	https://www.engenderhealth.org/
10	Family Health International 360 (FHI 360)	https://www.fhi360.org/countries/kenya
11	Hudara	https://hudara.org/learn-who-we-are/
12	Kenya Community Development Foundation (KCDF)	https://www.kcdf.or.ke/index.php/about-us
13	Kenyan Heart National Foundation	https://www.facebook.com/kenyanheart/
14	Kenya Voluntary Development Association (KVDA)	https://kvdakenya.org/
15	Komaza	https://www.tuko.co.ke/338191-list-ngos-kenya-contacts.html
16	Kickstart	https://kickstart.org/
17	Mercy Corps Kenya	https://www.mercycorps.org/where-we-work/kenya
18	StrongMinds	https://strongminds.org/
19	Ufadhili Trust	https://www.ufadhilitrust.org/
20	WellBoring	https://www.wellboring.org/

down by negative 1.6% in 2016, and up again by 3.6% in 2021, for examples.[31] The unemployment rate stood at 9.71% in 2021.[32] Nigeria took over from South Africa in 2013 as the biggest African economy and the

[31] https://data.worldbank.org/indicator/NY.GDP.MKTP.KD.ZG?locations=EG-KE-NG-ZA.

[32] Source for the unemployment rate: Egypt—Unemployment rate 2021 | Statista https://www.statista.com.

26th biggest in the world.[33] The top three most contributing sectors to Nigerian GDP in 2021 are:

1. Services 53.5%
2. Agriculture 25.8%
3. Industry 20.5%.[34]

Although popularly known as an oil producer (in fact the biggest in Africa), the oil sector's annual contribution to GDP actually ranges from only between 5–9% on annual basis.[35]

6.4.2 *Relevant National Laws/Regulations and ESG Reporting*

6.4.2.1 *Environmental*

The basis of environmental policy is contained in the 1999 Constitution of the Federal Republic of Nigeria. Pursuant to Section 20 of the Constitution, the State is empowered to protect and improve the conditions of the environment and safeguard the water, air and land, forest and wildlife of Nigeria. In addition to this, Section 2 of the Environmental Impact Assessment Act of 1992 (EIA Act) provides that the public and private sectors of the economy shall not undertake, embark on, or authorize projects or activities without prior consideration of the effects on the environment. The Federal Government has promulgated various laws and regulations to safeguard the Nigerian environment. The Federal Ministry of Environment (FME) administers and enforces environmental laws in Nigeria. It took over this function in 1999 from the Federal Environmental Protection Agency (FEPA), which was created under the FEPA Act. FEPA was absorbed and its functions taken over by the FME in 1999.

The Ministry of Environment has published several guidelines for the administration of the EIA Acts and the former FEPA procedures for

[33] AfDB, "Nigeria becomes Largest Economy in Africa with $509.9 Billion GDP," April 10, 2014, https://www.afdb.org/en/news-and-events/nigeria-becomes-largest-economy-in-africa-with-509-9-billion-gdp-12981#:~:text=Nigeria%20, last visited December 9, 2022.

[34] National Bureau of Statistics, *Nigerian Gross Domestic Product Q4 2021*, Table Ten, Abuja, February 2022.

[35] National Bureau of Statistics, ibid., p. 9; and https://www.statista.com/statistics/1165865/contribution-of-oil-sector-to-gdp-in-nigeria/.

evaluating environmental impact assessment reports. Other regulatory agencies with oversight over specific industries have also issued guidelines to regulate the impacts of these industries on the environment, such as the Environmental Guidelines and Standards for the Petroleum Industry in Nigeria (EGASPIN) in 2002, published by the Department of Petroleum Resources (DPR).

6.4.2.2 *Corporate Social Responsibility*

As in all other countries generally, CSR in Nigeria is influenced by the shareholder primacy doctrine or model. This means that companies' primary responsibility is to focus on shareholder profit maximization.[36] The Nigerian Companies and Allied Matters Act 1990 (CAMA) treats CSR as a voluntary approach to the needs of other parties, such as the community and larger society, by companies. According to some studies, this voluntary approach in Nigeria tended to have made directors believe that they have no legal obligation to anyone except shareholders and this believe has obstructed CSR practices and even, directly or indirectly, contributed to the commonly observed conflicts between companies and communities in Nigeria.[37] As a result of these conflicts and external pressures, the Nigerian government promulgated the Nigeria Extractive Industries Transparency Initiative (NEITI) Act in 2007. The aim was that the Act would promote accountability, transparency, and corporate social responsibility in manners that will lead to equitable social development, particularly through the petroleum industry. According to Ihugba (2012) the achievements of NEITI are mixed. On one hand it is yet to enforce its presence enough and achieve a significant mark in the areas of stakeholder engagement and social development; on the hand it has positively

[36] Discussed in Chapters 1 and 7.

[37] See, for instance, Ibukun Iyiola-Omisore, "Assessing the Relationship between the Nigerian Companies Act and Corporate Social Responsibility in Nigeria," Afronomicslaw, March 30, 2020, https://www.afronomicslaw.org/2020/03/30/assessing-the-relationship-between-the-nigerian-companies-act-and-corporate-social-responsibility-in-nigeria#:~:text=The%20CSR%20activities%20covered%20under,rights%2C%20environment%2C%20social%20business%20and, last visited January 5, 2020; Bethel U. Ihugba, "Compulsory Regulation of CSR: A Case Study of Nigeria," *Journal of Politics and Law*, Vol. 5, No. 2 (2012); and Amao Olufemi, "Corporate Social Responsibility, Multinational Corporations and the Law in Nigeria: Controlling Multinationals in Host States," *Journal of African Law*, Vol. 52 (2008): 83 at 96.

contributed to auditing oil company payments to the government.[38] The Extractive Industries Transparency initiative (2022) has concluded that although Nigeria is the largest producer of oil in Africa and 12th in the world, "the extractive sector has had a limited impact on socio-economic development in the past, in part due to weak governance and corruption. In this context, EITI implementation has helped to improve transparency of the sector's management and highlight areas in need of reform".[39]

6.4.2.3 Corporate Governance

Like environmental issues, the concept of corporate governance of some sort is not new in Africa.[40] Historically, it is generally and history embedded in company law, which can be traced back to the colonial era. Nigeria is a good example. The first Companies Act was passed in 1968 and replaced the one from the colonial period—the Companies Ordinance of 1922. In 1972, the Nigerian Enterprises Promotion Decree was promulgated and then repealed. A formal principle of corporate governance can be traced to the Companies and Allied Matters Act (CAMA) which was passed in 1990 and replaced the 1968 Companies Act, after the limitations of the latter Act were realized. The Act requires accounting and auditing of companies at set standards, regulation of the activities of companies, disclosure of shareholders, and definition of director's responsibilities. In addition to the CAMA, there were other company legislations, some repealed, some still in existence, such as the Banks and other Financial Institutions Act (BOFIA) 2004,[41] National Insurance Commission (NAICOM) Act 1997, Pension Commission (PENCOM) Act 2014, which replaced the Pension Reform Act No. 2, 2004, Financial Reporting Council (FRC) Act 2011, Nigerian Deposit Insurance Corporation (NDIC) Act 2006, Nigerian Communications (NCC) Act 2003, and Investment and Securities Act (ISA) 2007. Since the enacting of the CAMA, the corporate challenges around the world which are dealt with

[38] Ihugba, ibid.

[39] https://eiti.org/countries/nigeria; and also refer to NEITI, 2020 Oil and Gas Industry Report, published 2022.

[40] For the evolution of corporate governance in Nigeria, refer to, for instance, ABUAD Law Review, "An Analysis of Corporate Governance In Nigeria," *ABUAD Law Review*, Vol. 5 (2019/2020), https://djetlawyer.com/category/abuad-law-review/, last visited January 8, 2023.

[41] Replaced with the Banks and other Financial Institutions Act (BOFIA) 2020.

in greater detail in Chapters 1 and 7 brought the subject of corporate governance to the forefront, and countries started to examine corporate governance practices, reviewing old or passing new legislations and regulations. Distinct corporate governance codes came into vogue too. A similar trend has been found in all the four case studies in this book.

6.4.3 Relevant International Agreements/Treaties and ESG Reporting

Next, we present the outcome of our research on the international agreements/treaties which Nigeria might have already signed in each of the three ESG categories. As stated earlier, the objectives are to show whether the country already has international agreements (in addition to the national laws discussed above) for justifying and setting up comprehensive ESG reporting nationally, responding to the international calls for ESG reporting, and using these international agreements to reinforce the national cases for ESG disclosures. It is proven in all cases and ESG categories that the potential legal/regulatory frameworks and justifications exist and have been there for a very long time, although they were not in relation with, or intended for, comprehensive ESG as it is known today. The reader is referred to Appendix Table 6.11 for the detailed list of treaties.

6.4.4 The Biggest Locally Owned and Foreign Companies

Nigeria has the largest economy in Africa, as reported earlier. The population is also the largest in Africa, estimated to be around 200 million, and this population is highly attractive to investors and managers. Crude oil is the major generator of revenue for the government. The oil sector accounts for about 80% of the revenue and 90% of foreign exchange earnings, on average per annum. So it is not surprising that a number of the richest companies are mostly found in the oil and gas sector.[42] The

[42] For information on the top ten oil companies in Nigeria, you may refer to Peris Walubengo and Adrianna Simwa, "Top 10 Oil and Gas Companies in Nigeria and Their Websites," August 30, 2022, accessible at https://www.legit.ng/ask-legit/top/1130136-list-oil-gas-companies-nigeria-websites/, April 18, 2023.

wealthiest locally owned and foreign businesses operating in Nigeria are shown in Table 6.5.[43]

6.4.5 *The Nigerian Stock Exchange*

The Nigerian Stock Exchange (NSX) was founded in 1960 and it is the second biggest exchange in Africa, according to its market capitalization of about $67 billion. The exchange joined the United Nations Sustainable Stock Exchanges (SSE) in October 2013.

In July 2018, the Nigerian Stock Exchange became the first Exchange in Africa to list a sovereign green bond, with the listing by the Federal Debt Management Office (DMO) of a N10.69 billion[44] Series 1, 5-Year 13.48% Fixed Rate Bond due in 2022. In March 2019, Access Bank (the biggest bank in Nigeria) announced the first private sector Certified Climate Bond in Africa, with the issuance of a N15 billion[45] fixed rate senior unsecured green bond with a 15.5% fixed rate, maturing in 2024, on the NSX. Again in June 2019, NSX issued a N15 billion Series II Green Bond, with a coupon rate of 14.50% due to mature on 13 June 2026. In December 2021, the NSX reported that the value of Nigeria's green bonds market has grown to N55.52 billion between 2018 and 2021.[46]

The NSX launched its Sustainability Disclosure Guidelines in December 2018.[47] In addition to this, it became mandatory for all listed

[43] *Sources* Various. See, for instance, https://www.statista.com/statistics/1265989/leading-companies-in-nigeria-by-market-capitalization/, https://ngxgroup.com/exchange/trade/equities/listed-companies/ and https://ngxgroup.com/exchange/trade/equities/listed-companies/.

[44] The average spot exchange rate in July 2018: 360.58 to $1, https://www.exchangerates.org.uk/USD-NGN-15_12_2021-exchange-rate-history.html, November 4, 2022.

[45] The average spot exchange rate in March 2019: 360.12 to $1; The average spot exchange rate in June 2019: 347.7096, https://www.exchangerates.org.uk/USD-NGN-15_12_2021-exchange-rate-history.html, November 4, 2022.

[46] The Chief Executive Officer, Temi Popoola, at the Sustainable Finance Training 2021 hosted by NGX in collaboration with the International Finance Corporation (IFC) in December 2021. The average exchange rate in 2018: 361.29 NGN to $1; The average exchange rate in 2021: 403.58 NGN to $1, https://www.exchangerates.org.uk/USD-NGN-15_12_2021-exchange-rate-history.html, November 4, 2022.

[47] Sustainable Stock Exchanges, "Exchange in Focus: Nigeria Fulfills Commitment to Publish Guidance on Sustainability Reporting, Makes Mandatory," December 15,

Table 6.5 The biggest companies in Nigeria

Name	*Value of market capitalization (US$, millions)*[48]	*ESG/sustainability reporting*	*Origin*
1. Dangote Cement	11.203 billion	Yes https://www.dangotecement.com/sustainability-reporting/	Nigeria
2. MTN Nigeria	10.471 billion	Yes https://www.mtn.com/wp-content/uploads/2022/04/MTN-Group-FY-21-Sustainability-Report.pdf	South Africa
3. Airtel Africa	6.903 billion	Yes https://airtel.africa/assets/pdf/Airtel_Africa_Sustainability_Strategy_2021.pdf	India
3. BUA Cement	5.759 billion	Yes https://www.buacement.com/sustainability/	Nigeria
4. Nestle Nigeria	2.658 billion	Yes https://www.nestle-cwa.com/sites/g/files/pydnoa346/files/filefield_paths/SUSTAINABILITY%20POLICY%20OF%20NESTLE%CC%87%20NIGERIA%20PLC.pdf	Switzerland
5. BUA Foods	2.575 billion	Yes https://www.buafoodsplc.com/our-impact-esg/	Nigeria
6. Zenith Bank	1.691 billion	UN Global Compact https://www.unglobalcompact.org/participation/report/cop/create-and-submit/advanced/474572	Nigeria

[48] https://www.exchangerates.org.uk/Nigerian-Naira-to-Dollars-currency-conversion-page.html. As of the month of April 2023: 1 USD = 462.730588, Nigerian Naira.

Name	*Value of market capitalization (US$, millions)*	*ESG/sustainability reporting*	*Origin*
7. Guaranty Trust Holding Company	1.585 billion	CSR https://www.gtcoplc.com/how-we-give-back/csr-reports	Nigeria
8. FBN Holdings	1.070 billion	Yes https://www.fbnholdings.com/esg-information/	Nigeria
9. Stanbic IBTC Holdings	1.064 billion	Yes https://reporting.stanbicibtc.com/about-us/sustainability/	South African
10. Geregu Power PlC	715.697 million	Evidence unavailable https://www.geregupowerplc.com	Nigeria
11. Lafarge Cement Wapco	712.287 million	Yes https://www.lafarge.com.ng/sites/nigeria/files/2022-08/2021-sustainability-report.pdf	
12. Nigerian Breweries	676.416 million		
13. Access Bank Plc	618.684 million	Yes https://sustainability2020report.accessbankplc.com/report/	Nigeria
14. UBA	530.4332 million	Yes https://www.ubagroup.com/sustainability/	
15. Dangote Sugar	435.5868 million	Yes https://www.ubagroup.com/sustainability/	Nigeria
16. Union Bank of Nigeria	381.189 million	Yes https://www.unionbankng.com/blog/union-bank-releases-2021-citizenship-sustainability-and-innovation-report-highlights-commitment-to-environmental-and-social-impact/	Nigeria

(continued)

Table 6.5 (continued)

Name	*Value of market capitalization (US$, millions)*	*ESG/sustainability reporting*	*Origin*
17. EcoBank Transnational	380.6286 million	Yes https://ecobank.com/group/sustainability	
18. Okomu Oil Palm Co	355.7136 million	Evidence unavailable https://okomunigeria.com	Nigeria
19. Presco	265.1 million	Yes https://www.presco-plc.com/	Nigeria
20. Total Nigeria	163.438 million	Yes https://services.totalenergies.ng/	French

Sources https://www.statista.com/statistics/1265989/leading-companies-in-nigeria-by-market-capitalization/; https://www.statista.com/statistics/1265989/leading-companies-in-nigeria-by-market-capitalization/; and various

companies on its premium board to provide a sustainability report with effect from 1 January 2019. The Guidelines consist of four main areas: (1) the value proposition of sustainability; (2) integration of sustainability; (3) principles and core elements of a sustainability report; and (4) reporting requirements.

In March 2019, the NSX hosted a Sustainability Reporting Implementation Workshop in partnership with the Global Reporting Initiative as a way of helping listed companies adopt reporting principles in the NSX's Sustainability Disclosure Guidelines. On 30 May 2019, the Exchange launched the Facts Behind the Sustainability Report in order to promote the adoption of ESG practices and encourage responsible long-term approaches to investment. It provides a platform for listed companies to address stakeholders with in-depth analysis of their sustainability initiatives, as well as spotlight stakeholder engagements, materiality, standardization, and overall disclosures. According to the NSX, "better ESG reporting is key to strengthening capital markets and achieving a sustainable global economy. The Exchange is strategically positioned to influence the adoption of globally recognized sustainability standards by Nigerian businesses and we continue to highlight the importance of sustainable business practices in delivering value to our listed companies and investing public to support economic growth".[49]

6.4.6 Influential Non-Governmental Organizations (NGOs)

Nigeria has a very long list of NGOs. Below is a very short list of some of the most influential NGOs in the country (Table 6.6).[50]

2018: https://sseinitiative.org/all-news/nigeria-fulfills-commitment-to-publish-guidance-on-sustainability-reporting-makes-mandatory/.

[49] From statement by the Chief Executive Officer of the NSX, Mr. Oscar N. Onyema. Visit https://sseinitiative.org/all-news/nigeria-fulfills-commitment-to-publish-guidance-on-sustainability-reporting-makes-mandatory/, https://www.thefreelibrary.com/Dangote+Cement+Unveils+New+Sustainability+Report%2C+Adopts+GRI%2C+NSE...-a0587657967 and https://confiance.com.ng/esg-reporting-nse-unveils-facts-behind-the-sustainability-report/.

[50] For a longer list of more than 3000 NGOs, visit the Nigerian Network of NGOs at https://nnngo.org/list-of-ngos-in-nigeria/.

Table 6.6 Selected list of NGOs: Nigeria

	Name	*Website*
1	Action Aid for the Unemployed	https://nigeria.actionaid.org/
2	AIDS Prevention Initiative in Nigeria	https://www.devex.com/organizations/apin-public-health-initiatives-ltd-127587#:~:text=APIN%20is%20a%20leading%20Nigerian,diseases%20of%20public%20health%20significance
3	Amaudo UK	https://www.amaudo.org/
4	Bisi Alimi Foundation	https://www.bisialimifoundation.org/
5	Daisy's Home For Special Children Foundation	https://www.facebook.com/DaisysHomeNG/
6	Education as a Vaccine (EVA)	https://www.evanigeria.org/
7	GEANCO Foundation	https://www.geanco.org/
8	Global Peace Development	https://globalpeacedev.org/
9	HelpAge International	https://www.helpage.org/where-we-work/helpage-global-network-members-in-africa/
10	Lagos Food Bank Initiative	https://lagosfoodbank.org/
11	Land of Hope	https://landofhope.global/en/
12	MOCSOS: Mabel Oboh Centre for Save Our Stars	https://www.facebook.com/mabelobohcentresavestars/
13	Association of Urban and Rural Entrepreneurs of Nigeria	www.auare.org.ng
14	TASTE Nigeria	https://taste.org.uk/
15	The Irede Foundation	https://www.theiredefoundation.org/
16	Tony Elumelu Foundation	https://www.tonyelumelufoundation.org/
17	Total Care Unit	https://cleancooking.org/sector-directory/total-care-unit/
18	TY Danjuma Foundation	https://tydanjumafoundation.org/
19	Ven Dr. Josiah Alozie Foundation	https://www.wikidata.org/wiki/Q106573076
20	Yinka Shonibare Foundation	https://www.yinkashonibarefoundation.com/

6.5 South Africa

6.5.1 A Glimpse of the Economy

According to World Bank data, the South African economy grew from $65.65 billion in 1960 to $352.14 billion in 2021, based on GDP in constant 2015 US dollars.[51] However, over the years economic performance has slowed down considerably, from a GDP growth rate of 3.8% in 1960 to 7.9% in 1964, 4.4% in 1966, down by −0.1% in 1977, down by −1.1% in 1992, up 3.2% in 2011, down by −6.4 in 2020, and up by 4.9% in 2021, for examples.[52] The World Bank classifies South Africa as an upper-middle-income economy, and a newly industrialized country. Its economy is the second-largest in Africa, and the 34th-largest in the world. In terms of purchasing power parity, South Africa has the seventh-highest per capita income in Africa. However, poverty and inequality remain widespread, a common trait with all the case studies in the chapter. At 33.5% in 2021, the unemployment rate is extremely high.[53] The top three most contributing sectors to GDP are:

1. Services 62.7%
2. Industry 24.5%
3. Agriculture 2.4%.[54]

6.5.2 The Biggest Locally Owned and Foreign Companies

The wealthiest locally owned and foreign businesses operating in South Africa are shown in Table 6.7.

[51] https://data.worldbank.org/indicator/NY.GDP.MKTP.KD?locations=EG-KE-NG-ZA. All dollar figures throughout this book refer to United States dollars unless otherwise stated.

[52] https://data.worldbank.org/indicator/NY.GDP.MKTP.KD.ZG?locations=EG-KE-NG-ZA.

[53] Macrotrends, "South Africa Unemployment rate 1999–2022," https://www.macrotrends.net/countries/ZAF/south-africa/unemployment-rate, November 4, 2022.

[54] Statistica, "South Africa: Distribution of Gross Domestic Product (GDP) across Economic Sectors: 2011 to 2021," https://www.statista.com/statistics/371233/south-africa-gdp-distribution-across-economic-sectors/#:~:text=In%202021%2C%20agriculture%20had%20contributed,the%20total%20value%20added%2C%20respectively.

Table 6.7 The biggest companies in South Africa

Name	*Market value (US$, millions)*	*ESG/sustainability reporting*	*Origin*
1. Naspers	31.51 billion	Yes https://www.naspers.com/sustainability/esg-reports	South Africa
2. Firstrand	22.93 billion	Yes https://www.firstrand.co.za/media/investors/reports/FirstRand-report-to-society-2021.pdf	SA
3. Standard Bank Group	17.56 billion	Yes https://www.standardbank.com/static_file/StandardBankGroup/filedownloads/RTS/ESGReport2021.pdf	SA
4. MTN Group	14.30 billion	Yes https://www.mtn.com/wp-content/uploads/2022/04/MTN-Group-FY-21-Sustainability-Report.pdf	SA
5. Capitec Bank Holdings	13.66 billion	Yes https://www.capitecbank.co.za/globalassets/pages/documents-library/esg-documents/capitec-climate-related-financial-disclosure-2022.pdf	SA
6. Vodacom Group	12.12 billion	Yes https://vodacom-reports.co.za/integrated-reports/ir-2022/documents/Sustainability-report-2022.pdf	SA
7. SASOL	10.85 billion	Yes https://www.sasol.com/investor-centre/sustainability-reporting	SA
8. Impala Platinum Holdings	10.12 billion	Yes https://www.implats.co.za/esg-reporting.php	SA
9. Gold Fields	9.77 billion	Yes https://www.goldfields.com/sustainability-reporting.php	SA

Name	*Market value (US$, millions)*	*ESG/sustainability reporting*	*Origin*
10. Anglo American Platinum	7.70 billion	Yes https://www.angloamerican.com/~/media/Files/A/Anglo-American-Group/PLC/investors/annual-reporting/2022/aa-sustainability-report-full-2021.pdf	UK
11. Sibanye Stillwater	7.51 billion	Yes https://www.sibanyestillwater.com/sustainability/	SA
12. Sanlam	7.31 billion	Yes https://www.sanlam.com/sustainability	SA
13. Bid Corp	6.57 billion	Yes https://www.bidcorpgroup.com/sustainability-reporting.php	SA
14. Nedbank	6.49 billion	Yes https://www.nedbank.co.za/content/nedbank/desktop/gt/en/aboutus/green-and-caring/sustainability/Reports.html	SA
15. Discovery Limited	4.80 billion	Yes https://www.discovery.co.za/assets/discoverycoza/corporate/corporate-sustainability/2021/sustainability-report.pdf	SA
16. Pepkor	4.76 billion	Yes https://www.pepkor.co.za/sustainability/	SA
17. Remgro Limited	4.71 billion	Yes https://www.remgro.com/wp-content/uploads/2021/10/Sustainability-Report.pdf	SA
18. The Bidvest Group	4.48 billion	Yes https://www.bidvest.co.za/sustainability-overview.php	SA
19. Mediclinic International	4.39 billion	Yes https://www.mediclinic.com/en/sustainable-development.html	SA
20. Clicks Group	4.17 billion	Yes https://www.clicksgroup.co.za/wp-content/uploads/2022/08/sustainability-report-2021.pdf	SA

Sources https://www.value.today/headquarters/south-africa, https://naijaquest.com/biggest-companies-in-south-africa/ and https://companiesmarketcap.com/south-africa/largest-companies-in-south-africa-by-market-cap/

6.5.3 *Relevant National Laws/Regulations and ESG Reporting*

6.5.3.1 *Environmental*

The Constitution of the Republic of South Africa contains a Bill of Rights that includes several socio-economic and environmental rights that generally apply to the state as well as private persons (natural and juristic persons). Relevant fundamental rights include the rights to human dignity, freedom, and security and the right against slavery, servitude, and forced labour, the rights to life, healthcare, food, water, and social security, the right to an environment that is not harmful to human health or well-being (Section 24 (a)) and that is protected (Section 24 (b)) and the right of access to information. The National Environmental Management: Biodiversity Act, 2004 (Act No. 10 of 2004) established the South African National Biodiversity Institute (SANBI) in 2004. The Act expands the responsibilities beyond those of SANBI's forerunner, the National Botanical Institute, which was predominantly flora focused, and enabled SANBI to include responsibilities relating to the full spectrum of South Africa's biodiversity. SANBI carries out high quality research in the indigenous, naturalized, and alien flora of southern Africa and beyond. Its research programme covers conservation, biodiversity science, and climate change. SANBI is responsible for ensuring that biodiversity knowledge influences policy, management, and decision-making. The institute further operates environmental education programmes within its national botanical gardens and outreach greening to promote indigenous gardening at disadvantaged schools in surrounding communities.

6.5.3.2 *Corporate Social Responsibility*

The topics of Corporate Social Responsibility (CSR) and sustainability are well covered in what are known as the King Codes. Indeed, the country has the most advanced CSR outlook in Africa due to the domestic demands/pressure of CSR, the prevalence of mining in the economy, and the influence of corporate governance (Baskin 2006). Also, the Socially Responsible Investing (SRI) Index is shown to be the most advanced in the emerging economies (Baskin 2006).

6.5.3.3 *Corporate Governance*

South Africa is the pioneer of ESG, including corporate governance in Africa, and in fact the first Corporate Governance Code (King I) of the "King Reports" (discussed in more detail in Chapter 7) was completed in 1994, only two years after the first corporate governance code in the United Kingdom in 1992. This was before those of most other countries, including developed ones such as France (1995), Canada (1995) and the Netherlands Report (1995). The second Code in 2002 (i.e. King II), the third in 2009 (King III), and last revision in 2016 (King IV). The Reports are comprehensive and present well defined perspectives of corporate governance. However, investor protection still remains relatively low, because the external corporate regulatory environment is not strong enough, although more advanced than in other African countries.

6.5.4 *Relevant International Treaties/Agreements and ESG Reporting*

As in the cases of the other three countries in the regional case studies in this chapter, we investigated the existence and nature of international agreements which South Africa might have already signed in each of the three ESG categories. The objectives are to show whether the country already has international agreements (in addition to the national laws discussed above) for justifying and setting up comprehensive ESG reporting nationally, responding to the international calls for ESG reporting, and using these international agreements to reinforce the national cases for ESG disclosures. It is found in all cases and ESG categories that the potential legal/regulatory frameworks and justifications exist and have been there for a very long time, although they were not in relation with, or intended for, comprehensive ESG as it is known today. The reader is referred to Appendix Table 6.12 for more details.

6.5.5 *The Johannesburg Stock Exchange*

The Johannesburg Stock Exchange (JSE) is the biggest stock exchange and the second oldest in Africa. With market capitalization of $1.36 trillion, it also ranks 13th in the world.[55] It is the most advanced in Africa in its sustainability measures. Since South Africa (including the JSE) is discussed in great detail in Chapter 7, the reader is referred to that chapter for more information.

[55] For global rankings of stock exchanges see the Globaleconomy.com, "Stock Market Capitalization, in Dollars—Country Rankings," https://www.theglobaleconomy.com/rankings/stock_market_capitalization_dollars/,/.

6.5.6 *Influential Non-Governmental Organizations (NGOs)*

There are numerous non-government organizations (NGOs) in South Africa helping those in need.[56] These organizations formed the Southern African NGO Network (SANGONeT) in 1987. Since then, the network has developed into a civil society group that is historically linked to the social and political changes experienced in South Africa before and after democracy was instituted in 1994. Despite being part of a network, the NGOs in South Africa also work independently. The NGOs in South Africa that are working to make a difference on issues of relevance to ESG include (Table 6.8).[57]

[56] GlobalGiving Atlas has a database of 255,745 NGOs in South Africa. https://www.globalgiving.org/atlas/country/south-africa/.

[57] *Sources* https://borgenproject.org/ngos-in-south-africa/, https://www.devex.com/news/top-global-development-ngos-in-south-africa-a-primer-77094 and https://www.wango.org/resources.aspx?section=ngodir&sub=region&regionID=18&col=BFB07D.

Table 6.8 Selected list of NGOs: South Africa

	Name	*Website*
1	Sustainability Initiative of South Africa (SIZA)	https://siza.co.za/
2	AIDS Foundation of South Africa	https://www.aids.org.za/
3	Association for the Physically Disabled—Greater Johannesburg	https://apdjhb.co.za/
4	CHOSA South Africa	https://www.chosa.org/
5	Devex	https://www.devex.com/countries/south-africa-1160
6	Empowerment for African Sustainable Development	https://gyan.tigweb.org/members/
7	Global Youth Action Network	https://gyan.tigweb.org/members/
8	Kickstart	https://kickstart.org/
9	MIET Africa	https://mietafrica.org/
10	mothers2mothers	https://m2m.org/
11	Mvula Trust	https://themvulatrust.org.za/
13	One	https://www.one.org/international/
14	READ Educational Trust	https://www.read.org.za/
15	Rural Community Initiative	https://www.rd.usda.gov/programs-services/community-facilities/rural-community-development-initiative-grants
16	Save the Children South Africa	https://www.rd.usda.gov/programs-services/community-facilities/rural-community-development-initiative-grants
17	The Future Kids	thefuturekids.co.za
18	The Rural Women's Movement	https://www.wocan.org/organization/rural-womens-movement/
19	The Viva Foundation of South Africa	https://www.vivafoundation.life/
20	Wildlife and Environment Society of South Africa	https://wessa.org.za/

Appendix

See Tables 6.9, 6.10, 6.11, and 6.12.

Table 6.9 Egypt's relevant international agreements/treaties

Environmental (Egypt)
• African Convention on the Conservation of Nature and Natural Resources, 1968 • Agreement on the Application of Sanitary and Phytosanitary Measures, 1995 • Agreement on the Conservation of African-Eurasian Migratory Waterbirds, 1995, 1999 • Agreement on the Conservation of Cetaceans of the Black Sea, Mediterranean Sea and Contiguous Atlantic Area Ballast Water Management Convention, 1996 • Bamako Convention 1991 • Barcelona Convention 1976 • Convention on Biological Diversity, 1992, 1993 • Cartagena Protocol on Biosafety, 2000, 2003 • International Convention for the Conservation of Atlantic Tunas, 1966, 1969 • Convention on Assistance in the Case of a Nuclear Accident or Radiological Emergency, 1986 • Convention on Early Notification of a Nuclear Accident, 1986 • Convention on the Conservation of Migratory Species of Wild Animals, 1979, 1983 • Convention on the Limitation Period in the International Sale of Goods, 1974, 1988 • Environmental Modification Convention, 1994, 1996 • International Convention on Oil Pollution Preparedness, Response and Co-operation, 1990 • International Convention on the Control of Harmful Anti-fouling Systems on Ships, 2001, 2008 • International Convention Relating to Intervention on the High Seas in Cases of Oil Pollution Casualties, 1969 • International Convention on Civil Liability for Bunker Oil Pollution Damage, 2008 • International Treaty on Plant Genetic Resources for Food and Agriculture, 2001 • International Tropical Timber Agreement, 2006 • International Tropical Timber Agreement, 1983,1994, 2006

Environmental (Egypt)
• Kyoto Protocol, 1992 • London Convention on the Prevention of Marine Pollution by Dumping of Wastes and Other Matter, 1972 • MARPOL 1973 • Montreal Protocol 1987 Paris Agreement, 2015 • Partial Nuclear Test Ban Treaty, 1963 • Radiation Protection Convention, 1960 • Ramsar Convention 1971 • Section 123 Agreement, 1954[58] • Statute of the International Renewable Energy Agency, 2009 • Stockholm Convention on Persistent Organic Pollutants, 2001, 2004 • Treaty on the Non-Proliferation of Nuclear Weapons, 1968 • United Nations Convention on the Law of the Sea 1982 • United Nations Convention to Combat Desertification, 1994, 1996[59] • United Nations Framework Convention on Climate Change, 1994 • Vienna Convention for the Protection of the Ozone Layer, 1985, 1988 • Vienna Convention on Civil Liability for Nuclear Damage, 1963, 1977, 1997 • Working Environment (Air Pollution, Noise and Vibration) Convention, 1977

(continued)

58 Also relevant to Governance.

59 Sara Refaat, *Corporate social responsibility in Egypt: A study on the current practice, challenges*, and potentials, American University in Cairo, AUC Knowledge Fountain, 2014, https://core.ac.uk/download/pdf/333724443.pdf, last visited 16 December 2022.

Table 6.9 (continued)

Social (Egypt)
• African Youth Charter, 2009 • Bamako Convention, 1991, 1998 • Committee Against Torture, 1987 • Constitution of the International Labour Organization, 1919 • Convention Against Discrimination in Education, 1962 • Convention for the Safeguarding of the Intangible Cultural Heritage • Convention on the Rights of the Child, 1989 • Continuity of Employment (Seafarers) Convention, 1976 • Convention Governing the Specific Aspects of Refugee Problems in Africa, 1969 • Convention on the Elimination of All Forms of Discrimination Against Women, 1979 • Convention on the Political Rights of Women • Convention on the Protection and Promotion of the Diversity of Cultural Expressions, 2005, 2007 • Convention on the Protection of the Underwater Cultural Heritage, 2001 • Convention on the Rights of Persons with Disabilities, 2007, 2008 • Dock Work Convention, 1973 • Equality of Treatment (Social Security) Convention, 1962 • Final Articles Revision Convention, 1961 • Hague Divorce Convention • Human Resources Development Convention, 1975 • International Convention Against Doping in Sport, 2005 • International Covenant on Civil and Political Rights, 1954, 1966, 1976 • International Covenant on Economic, Social and Cultural Rights, 1966, 1976 • International Convention on the Elimination of All Forms of Racial Discrimination • Labour Administration Convention, 1978 • Labour Inspection (Agriculture) Convention, 1969

Social (Egypt)
• Medical Examination (Seafarers) Convention, 1946 • Merchant Shipping (Minimum Standards) Convention, 1976 • Migrant Workers Convention • Minimum Age Convention, 1973 • Minimum Wage Fixing Convention 1970 • Nursing Personnel Convention, 1977 • Occupational Cancer Convention, 1974 • Occupational Safety and Health (Dock Work) Convention, 1979 • Optional Protocol on the Involvement of Children in Armed Conflict, 2000, 2002 • Optional Protocol on the Sale of Children, Child Prostitution and Child Pornography, 2000, 2002 • Placing of Seamen Convention, 1920 • Prevention of Accidents (Seafarers) Convention, 1970 • Protocol Against the Smuggling of Migrants by Land, Sea and Air • Protocol Amending the Single Convention on Narcotic Drugs • Protocol Relating to the Status of Refugees, 1967 • Repatriation of Seafarers Convention (Revised), 1987 • Repatriation of Seamen Convention, 1926 • Rescue Agreement, 1967 • Seafarers' Pensions Convention, 1946 • Seamen's Articles of Agreement Convention, 1926 • Shipowners' Liability (Sick and Injured Seamen) Convention, 1936

(continued)

Table 6.9 (continued)

Social (Egypt)
• Sickness Insurance (Sea) Convention, 1936 • Single Convention on Narcotic Drugs • SOLAS Convention, 1960, 1974 • STCW Convention, 1978 • Tripartite Consultation (International Labour Standards) Convention, 1976 • UNESCO Convention on the Means of Prohibiting and Preventing the Illicit Import, Export and Transfer of Ownership of Cultural Property, 1970 • United Nations Convention Against Corruption, 2003 • United Nations Convention Against Illicit Traffic in Narcotic Drugs and Psychotropic Substances, 1988, 1990 • Vocational Rehabilitation and Employment (Disabled Persons) Convention, 1983 • Workers' Representatives Convention, 1971 • Working Environment (Air Pollution, Noise and Vibration) Convention, 1977 • World Heritage Convention, 1973 • Worst Forms of Child Labour Convention, 1999

Governance (Egypt)
• Agreement relating to the International Telecommunications Satellite Organization, 1971, 1973 • Arms Trade Treaty, 2013, 2014 • Convention on Certain Conventional Weapons, 1980, 1983 • Convention for the Mutual Recognition of Inspections in respect of the Manufacture of Pharmaceutical Products, 1970 • Convention on Cluster Munitions, 2008 • Convention on Mutual Administrative Assistance in Tax Matters, 2016, 2017 • Convention on Psychotropic Substances, 1971 • Convention respecting the Limitation of the Employment of Force for Recovery of Contract Debts, 1907 • Cotonou Agreement, 2000 • Final Articles Revision Convention, 1961 • General Agreement on Trade in Services, 1995 • Hague Conventions of 1899 and 1907 • International Grains Agreement, 1994, 1995 • International Convention on the Harmonization of Frontier Controls of Goods, 1982 • International Convention on the Harmonized Commodity Description and Coding System, 1988 • International Sugar Agreement, 1992 • Lusaka Accords, 1984

(continued)

Table 6.9 (continued)

Governance (Egypt)
• Maputo Protocol, 2003, 2005[60] • Protocol on Incendiary Weapons, 1980, 1983 • Protocol on Mines, Booby-Traps and Other Devices, 1983, 1996, 1998 • Nagoya Protocol, 2010 • Nairobi Convention, 1985 • Nkomati Accord, 1984 • Ottawa Treaty, 1997, 1999 • Patent Cooperation Treaty, 1970 • Protocol amending the Single Convention on Narcotic Drugs • Protocol on Blinding Laser Weapons, 1980, 1995 • Ramsar Convention • Registration Convention • Rome Statute • Rotterdam Convention • International Convention for Safe Containers • Section 123 Agreement, 1954 • Settlement Proposal • Single Convention on Narcotic Drugs

Source Various—Wikipedia, authors, etc.

[60] Also relevant to Social.

Table 6.10 Kenya's International agreements/treaties

Environmental (Kenya)
• African Convention on the Conservation of Nature and Natural Resources, 1968, 1969 • African Nuclear-Weapon-Free Zone Treaty, 1996, 2009 • Agreement for Establishment of the Global Crop Diversity Trust, 2004 • Agreement for the Establishment of the Indian Ocean Tuna Commission, 1993, 1996 • Agreement on Port State Measures to Prevent, Deter and Eliminate Illegal, Unreported and Unregulated Fishing, 2009 • Agreement on the Conservation of African-Eurasian Migratory Waterbirds, 1995 • Ballast Water Management Convention, 2004, 2017 • Biological Weapons Convention, 1992 • Brussels Collision Convention, 1910[61] • Cartagena Protocol on Biosafety, 2000, 2003 • Chemical Weapons Convention, 1993, 1997[62] • CITES, 1973 • Comprehensive Nuclear-Test-Ban Treaty, 1996 • Convention for the Protection of New Varieties of Plants, 1998 • Convention on Biological Diversity, 1992, 1993 • Convention on Fishing and Conservation of the Living Resources of the High Seas, 1958 • Convention on the Conservation of Migratory Species of Wild Animals, 1979, 1983 • Convention on the High Seas, 2001, 2008

(continued)

[61] Also relevant to Social (health and safety) and Governance (regulating international transportation).
[62] Also relevant to Governance.

Table 6.10 (continued)

Environmental (Kenya)
• Convention on the Marking of Plastic Explosives, 1991 • Convention on the Physical Protection of Nuclear Material, 1979 • Geneva Protocol, 1925[63] • International Convention for Safe Containers,[64] 1972, 1997, 2002 • International Convention for the Protection of New Varieties of Plants, 1998 • International Convention for the Regulation of Whaling, 1946 • International Convention on Civil Liability for Bunker Oil Pollution Damage, 2008 • International Convention on Civil Liability for Oil Pollution Damage, 1969, 1992 • International Convention on Oil Pollution Preparedness, Response and Co-operation, 1990 • International Convention on the Control of Harmful Anti-fouling Systems on Ships, 2001, 2008 • International Convention on the Establishment of an International Fund for Compensation for Oil Pollution Damage, 1992, 2006 • International Plant Protection Convention, 1951 • International Regulations for Preventing Collisions at Sea, 1972 • Kyoto Protocol, 1992 • London Convention on the Prevention of Marine Pollution by Dumping of Wastes and Other Matter, 1972, 1975 • MARPOL 1973, 1978 • Memorandum of Understanding on the Conservation and Management of Marine Turtles and their Habitats of the Indian Ocean and South-East Asia, 2001 • Memorandum of Understanding on the Conservation of Migratory Sharks, 2010 • Nagoya Protocol, 2010 • Nairobi Convention, 1985[65]

[63] Also relevant to Social and Governance.

[64] Also relevant to Governance.

[65] Of particular importance and a model of public–private partnership in ESG as in the South African fund.

Environmental (Kenya)
• Nairobi International Convention on the Removal of Wrecks, 2007, 2015 • Nuclear Terrorism Convention,[66] 2005, 2007 • Ottawa Treaty, 1997 • Outer Space Treaty, 1967 • Paris Agreement, 2015 • Partial Nuclear Test Ban Treaty, 1963 • Protocol for the Suppression of Unlawful Acts against the Safety of Fixed Platforms Located on the Continental Shelf, 1988, 1992 • Ramsar Convention, 1971 • Rome Convention on Damage Caused by Foreign Aircraft to Third Parties on the Surface, 1952 • Rotterdam Convention, 1998 • Statute of the International Atomic Energy Agency, 1956, 1957 • Statutes of the International Centre for Genetic Engineering and Biotechnology, 1983, 1994 • Statute of the International Renewable Energy Agency, 2009, 2010 • Treaty on the Non-Proliferation of Nuclear Weapons, 1968, 1970 • United Nations Convention on the Law of the Sea, 1982, 1994 • United Nations Convention to Combat Desertification, 1995, 1996

(continued)

[66] Also relevant to governance.

Table 6.10 (continued)

Social (Kenya)
• Abolition of Forced Labour Convention, 1957, 1959 • African Charter on Human and Peoples' Rights, 1979, 1981, 1986 • African Charter on the Rights and Welfare of the Child, 1990, 1999 • African Union Convention on Preventing and Combating Corruption, 2003 • African Youth Charter, 2006, 2009 • Agreement on the Importation of Educational, Scientific and Cultural Materials, 1950, 1952 Arusha Agreement, 1969, 1971 • Committee Against Torture, 1987 • Constitution of the International Labour Organization, 1919 • Constitution of the International Organization for Migration, 1951 • Convention Concerning Statistics of Wages and Hours of Work, 1938 • Convention for the Safeguarding of the Intangible Cultural Heritage, 2003, 2006 • Convention for the Suppression of Unlawful Acts against the Safety of Civil Aviation, 1971, 1973 • Convention for the Suppression of Unlawful Acts against the Safety of Maritime Navigation, 1988, 1992 • Convention on the Non-Applicability of Statutory Limitations to War Crimes and Crimes Against Humanity, 1968, 1970 • Convention on the Protection and Promotion of the Diversity of Cultural Expressions, 2005, 2007 • Convention on the Rights of Persons with Disabilities, 2007, 2008 • Convention on the Rights of the Child, 1989, 1990 • Convention Relating to the Status of Refugees, 1951, 1954 • Discrimination (Employment and Occupation) Convention, 1958, 1960

Social (Kenya)
• Dock Work Convention, 1973 • Employment Service Convention, 1948 • Equal Remuneration Convention, 1951, 1953 • Equality of Treatment (Accident Compensation) Convention, 1925 • Equality of Treatment (Social Security) Convention, 1962 • First Geneva Convention, 1864 • Fourth Geneva Convention, 1949 • Georgetown Agreement, 1975 • Hague Adoption Convention, 1999, 1995 • Hague Hijacking Convention, 1970, 1971 • Holidays with Pay Convention (Revised), 1970 • Human Resources Development Convention, 1975 • International Covenant on Civil and Political Rights, 1966, 1976 • International Covenant on Economic, Social and Cultural Rights, 1966, 1976 • International Convention on Maritime Search and Rescue, 1979, 1985 • International Convention on the Elimination of All Forms of Racial Discrimination, 1965, 1969 • Labour Clauses (Public Contracts) Convention, 1949 • Labour Inspection (Agriculture) Convention, 1969 • Labour Inspection Convention, 1947 • Maritime Labour Convention, 2006, 2013

(continued)

Table 6.10 (continued)

Social (Kenya)
• Maputo Protocol, 2003, 2005 • Medical Examination of Young Persons (Sea) Convention, 1921 • Migrant Workers (Supplementary Provisions) Convention, 1975, 1978 • Migration for Employment Convention (Revised), 1949 • Minimum Age Convention, 1973 • Minimum Wage Fixing Convention 1970 • Minimum Wage Fixing Machinery (Agriculture) Convention, 1951 • Minimum Wage-Fixing Machinery Convention, 1928 • Montreal Convention, 1999 • Nakuru County Peace Accord, 2012 • Nursing Personnel Convention, 1977 • Optional Protocol on the Involvement of Children in Armed Conflict, 2000, 2002 • Paid Educational Leave Convention, 1974 • Prevention of Accidents (Seafarers) Convention, 1970

Social (Kenya)
• Protection against Accidents (Dockers) Convention (Revised), 1932 • Protocol to the African Charter on Human and Peoples' Rights on the establishment of an African Court on Human and Peoples' Rights, 1998, 2004 • Protocol Relating to the Status of Refugees, 1967 • Right of Association (Agriculture) Convention, 1921, 1923 • Right to Organise and Collective Bargaining Convention, 1949 • Rural Workers' Organisations Convention, 1975 • Seafarers' Annual Leave with Pay Convention, 1976 • Second Geneva Convention, 1906 • Third Geneva Convention, 1929 • Treaty of Bern, 1874 • United Nations Convention Against Illicit Traffic in Narcotic Drugs and Psychotropic Substances, 1988, 1990 • United Nations Convention Against Corruption, 2003, 2005
Governance (Kenya)
• Agreement on Agriculture, 1995 • Agreement on Trade-Related Investment Measures, 1994, 1995 • Arusha Agreement, 1993

(continued)

Table 6.10 (continued)

Governance (Kenya)
• Chicago Convention on International Civil Aviation, 1995 • Convention on the Continental Shelf, 1958 • Cotonou Agreement, 2000, 2003, 2005, 2010 • First Geneva Convention, 1864 • Fourth Geneva Convention, 1949 • Georgetown Agreement, 1975 • Hague Conventions of 1899 and 1907 • International Grains Agreement, 1995 • International Coffee Agreement, 1962, 2007 • International Convention for the Suppression of Counterfeiting Currency, 1929 • Maputo Protocol, 2003, 2005 • MARPOL, 1973, 1978 • Marrakesh Agreement, 1994, 1995 • Nakuru County Peace Accord, 2012

Governance (Kenya)
• Ottawa Treaty, 1997, 1999 • Paris Convention for the Protection of Industrial Property, 1883, 1884 • Protocol I, 1977 • Protocol II, 1977 • Protocol III, 2005 • Ramsar Convention, 1971 • Rome Statute, 1998, 2002 • Rotterdam Convention, 1998 • Second Geneva Convention, 1906 • Third Geneva Convention, 1929 • United Nations Convention Against Illicit Traffic in Narcotic Drugs and Psychotropic Substances, 1988, 1990

Source Various—Wikipedia, authors, etc.

Table 6.11 Nigeria's Relevant international agreements/treaties

Environmental (Nigeria)
• African Nuclear-Weapon-Free Zone Treaty, 1996, 2009 • Agreement on the Application of Sanitary and Phytosanitary Measures, 1995 • Agreement on the Conservation of African-Eurasian Migratory Waterbirds, 1999 • Agreement on the Conservation of Gorillas and Their Habitats, 2007, 2008 • Ballast Water Management Convention, 2004, 2017 • Biological Weapons Convention, 1972, 1975 • Brussels Collision Convention, 1910, 1913 • Cartagena Protocol on Biosafety, 2000, 2003 • Chemical Weapons Convention,[67] 1993, 1997 • CITES, 1973, 1975 • Comprehensive Nuclear-Test-Ban Treaty,[68] 1996 • Convention on Assistance in the Case of a Nuclear Accident or Radiological Emergency, 1986, 1987 • Convention on Biological Diversity, 1992, 1993 • Convention on Early Notification of a Nuclear Accident, 1986 • Convention on Fishing and Conservation of the Living Resources of the High Seas, 1958, 1966 • Convention on Nuclear Safety, 1994, 1996 • Convention on the Conservation of Migratory Species of Wild Animals, 1979, 1983 • Convention on the High Seas, 1958, 1962 • Convention on the Law of the Non-Navigational Uses of International Watercourses, 1997 • Convention on the Marking of Plastic Explosives, 1991, 1998

67 Relevant to Governance also.

68 Relevant to Governance also.

Environmental (Nigeria)
• Convention on the Physical Protection of Nuclear Material, 1979, 1987 • Geneva Protocol, 1925 • International Convention for the Conservation of Atlantic Tunas, 1966 • International Convention on the Control of Harmful Anti-fouling Systems on Ships, 2001, 2008 International Convention on Civil Liability for Oil Pollution Damage, 1969, 1992 • International Convention on Oil Pollution Preparedness, Response and Co-operation, 1990, 1995 • International Convention Relating to Intervention on the High Seas in Cases of Oil Pollution Casualties, 1969, 1975 • International Convention on the Establishment of an International Fund for Compensation for Oil Pollution Damage, 1992, 2006 • International Plant Protection Convention, 1951, 1952 • International Regulations for Preventing Collisions at Sea, 1972 • Joint Convention on the Safety of Spent Fuel Management and on the Safety of Radioactive Waste Management, 1997, 2001 • Kyoto Protocol, 1997, 2005 • London Convention on the Prevention of Marine Pollution by Dumping of Wastes and Other Matter, 1972 • MARPOL, 1973, 1978 • Memorandum of Understanding concerning Conservation Measures for Marine Turtles of the Atlantic Coast of Africa, 1999, 2008 • Montreal Protocol, 1987, 1989 • Nairobi International Convention on the Removal of Wrecks, 2007, 2015 • Nuclear Terrorism Convention,[69] 2005, 2007 • Ottawa Treaty, 1997[70] • Outer Space Treaty, 1967 • Paris Agreement, 2016 • Partial Nuclear Test Ban Treaty, 1963

(continued)

[69] Also relevant to Governance.

[70] Also relevant to Governance.

Table 6.11 (continued)

Environmental (Nigeria)
• Statute of the International Atomic Energy Agency, 1957 • Statute of the International Renewable Energy Agency, 2009, 2010 • Statutes of the International Centre for Genetic Engineering and Biotechnology, 1983, 1994 • Treaty on the Non-Proliferation of Nuclear Weapons, 1968, 1970 • United Nations Convention on the Law of the Sea, 1982, 1994 • United Nations Convention to Combat Desertification, 1994, 1995, 1996
Social (Nigeria)
• Abolition of Forced Labour Convention, 1957, 1959 • Accommodation of Crews (Supplementary Provisions) Convention, 1970 • ADR (treaty), 1957, 1968, 1985, 2011, 2015 • African Charter on Human and Peoples' Rights, 1979, 1981, 1986, 1987, 1999, 2004 • African Charter on the Rights and Welfare of the Child, 1990, 1999 • African Union Convention on Preventing and Combating Corruption, 2003 • African Youth Charter, 2006, 2009

Social (Nigeria)
• Agreement for the Suppression of the Circulation of Obscene Publications, 1910, 1949, 1950 • Agreement on the Importation of Educational, Scientific and Cultural Materials, 1950, 1952 • Convention on the Rights of the Child, 1989, 1990 • Committee Against Torture, 1987 • Constitution of the International Labour Organization, 1919 • Convention Against Discrimination in Education, 1960, 1962 • Convention for the Suppression of the Circulation of and Traffic in Obscene Publications, 1923, 1924, 1947 • Convention on Mutual Administrative Assistance in Tax Matters, 1988, 1995 • Convention on the Non-Applicability of Statutory Limitations to War Crimes and Crimes Against Humanity, 1968, 1970 • Convention on the Political Rights of Women, 1953, 1954 • Convention on the Protection and Promotion of the Diversity of Cultural Expressions, 2005, 2007 • Convention on the Protection of the Underwater Cultural Heritage, 2001, 2009 • Convention on the Rights of Persons with Disabilities, 2007, 2008

(continued)

Table 6.11 (continued)

Social (Nigeria)
• Discrimination (Employment and Occupation) Convention, 1958, 1960 • Dock Work Convention, 1973 • Employment Service Convention, 1948 • Equal Remuneration Convention, 1951, 1953 • Equality of Treatment (Accident Compensation) Convention, 1925 • Final Articles Revision Convention, 1961 • Freedom of Association and Protection of the Right to Organise Convention, 1948, 1950 • Geneva Declaration on Armed Violence and Development, 2006 • Hague Convention for the Protection of Cultural Property in the Event of Armed Conflict, 1954, 1956 • International Agreement for the suppression of the White Slave Traffic, 1904, 1905, 1910, 1920 • International Convention Against Doping in Sport, 2005, 2007 • International Covenant on Civil and Political Rights, 1966, 1976 • International Covenant on Economic, Social and Cultural Rights, 1966, 1976 • International Convention for the Protection of All Persons from Enforced Disappearance, 2007, 2010 • Labour Clauses (Public Contracts) Convention, 1949 • Labour Inspection (Seafarers) Convention, 1996 • Labour Inspection Convention, 1947 • Lagos Treaty of Cession, 1961 • Maputo Protocol, 2003 • Maritime Labour Convention, 2006, 2013

Social (Nigeria)
• Marrakesh Agreement, 1994 • Medical Examination of Young Persons (Sea) Convention, 1921 • Migrant Workers Convention, 1990, 2003 • Migration for Employment Convention (Revised), 1949 • Minimum Age Convention, 1973 • Minimum Wage-Fixing Machinery Convention, 1928 • Montreal Protocol, 1987 • Optional Protocol on the Involvement of Children in Armed Conflict, 2000, 2002 • Optional Protocol on the Sale of Children, Child Prostitution and Child Pornography, 2000, 2002 • Optional Protocol to the Convention on the Rights of Persons with Disabilities, 2005, 2007 Occupational Safety and Health Convention, 1981 • Prevention of Accidents (Seafarers) Convention, 1970 • Protection against Accidents (Dockers) Convention (Revised), 1932 • Protection of Wages Convention, 1949 • Protocol Bringing under International Control Drugs outside the Scope of the Convention of 13 July 1931 for Limiting the Manufacture and Regulating the Distribution of Narcotic Drugs, 1948, 1949 • Protocol for the Suppression of Unlawful Acts of Violence at Airports, 1971, 1973

(continued)

Table 6.11 (continued)

Social (Nigeria)
• Protocol to the African Charter on Human and Peoples' Rights on the establishment of an African Court on Human and Peoples' Rights, 1998, 2004, 2009 • United Nations Convention Against Corruption, 2003, 2005 • United Nations Convention Against Illicit Traffic in Narcotic Drugs and Psychotropic Substances, 1988, 1990
Governance (Nigeria)
• Agreement on Agriculture, 1995 • Agreement on Trade-Related Investment Measures, 1994, 1995 • Aircraft Protocol to the Cape Town Treaty, 2001, 2006 • Arms Trade Treaty, 2013, 2014 • Cotonou Agreement, 2000, 2003, 2005, 2010 • Fourth Geneva Convention, 1949 • General Agreement on Trade in Services First Geneva Convention, 1864 • Geneva Declaration on Armed Violence and Development, 2006, 2008, 2011 • Geneva Protocol, 1925

Governance (Nigeria)
• Georgetown Agreement, 1975 • Greentree Agreement, 2006 • Hague Conventions of 1899 and 1907 • Hague Protocol, 1955, 1963 • Hamburg Rules Hostages Convention, 1979, 1983 • International Cocoa Agreement (2001), 1973, 2010, 2012 • Maputo Protocol, 2003, 2005 • Maroua Declaration, 1975 • Marrakesh Agreement, 1994 • Montreal Convention, 1999, 2003 • Ottawa Treaty, 1997, 1999 • Protocol I, 1977 • Protocol II, 1977 • Second Geneva Convention, 1906 • Third Geneva Convention, 1929 • Treaty of Lagos, 1975 • UNIDROIT Statute, 1993

Source Various—Wikipedia, authors, etc.

Table 6.12 South Africa's Relevant international treaties/agreements

Environmental (South Africa)
• African Nuclear-Weapon-Free Zone Treaty, 1996, 2009 • Agreed Measures for the Conservation of Antarctic Fauna and Flora, 1964 • Agreement on Port State Measures to Prevent, Deter and Eliminate Illegal, Unreported and Unregulated Fishing, 2009 • Agreement for the Establishment of the Indian Ocean Tuna Commission, 1993, 1996 • Agreement on the Application of Sanitary and Phytosanitary Measures, 1995 • Agreement on the Conservation of African-Eurasian Migratory Waterbirds, 1995, 1999 • Agreement on the Conservation of Albatrosses and Petrels, 2001 • Antarctic Treaty System, 1959 • Ballast Water Management Convention, 2004, 2017 • Biological Weapons Convention, 1972 • Cape Town Treaty, 2001, 2006 • Cartagena Protocol on Biosafety, 2000, 2003 • Chemical Weapons Convention, 1993, 1997 • CITES, 1973, 1975 • Commission for the Conservation of Southern Bluefin Tuna, 1994 • Convention for the Conservation of Antarctic Marine Living Resources, 1980, 1982 • Convention for the Conservation of Antarctic Seals, 1972 • Convention for the Protection of New Varieties of Plants, 1961, 1991 • Convention on Assistance in the Case of a Nuclear Accident or Radiological Emergency, 1986, 1987 • Convention on Biological Diversity, 1992, 1993

Environmental (South Africa)
• Convention on Cluster Munitions, 2008, 2010 • Convention on Early Notification of a Nuclear Accident, 1986 • Convention on Fishing and Conservation of the Living Resources of the High Seas, 1958, 1966 • Convention on the Conservation and Management of Fishery Resources in the South East Atlantic Ocean, 1971 • Convention on the Conservation of Migratory Species of Wild Animals, 1979, 1983 • Convention on the High Seas, 1958 • Convention on the Law of the Non-Navigational Uses of International Watercourses, 1997[71] • Convention on the Marking of Plastic Explosives, 1991, 1998 • Convention on the Physical Protection of Nuclear Material, 1979, 1987 • Convention on the Territorial Sea and the Contiguous Zone, 1958 • Doha Amendment to the Kyoto Protocol, 2012 • Hague Protocol, 1955 • International Convention for the Conservation of Atlantic Tunas, 1966 • International Convention for the Protection of New Varieties of Plants, 1961 and revised in 1972, 1978 and 1991 • International Convention on Civil Liability for Oil Pollution Damage, 1969, 1992 • International Convention on the Control of Harmful Anti-fouling Systems on Ships, 2001, 2008 • International Convention on the Establishment of an International Fund for Compensation for Oil Pollution Damage, 1992, 2006 • International Convention Relating to Intervention on the High Seas in Cases of Oil Pollution Casualties, 1969, 1975, 2002 • International Plant Protection Convention, 1951, 1952 • International Regulations for Preventing Collisions at Sea, 1972 • Joint Convention on the Safety of Spent Fuel Management and on the Safety of Radioactive Waste Management, 1997, 2001

(continued)

[71] Also relevant for Governance.

Table 6.12 (continued)

Environmental (South Africa)
K • Kyoto Protocol, 1997 • London Convention on the Prevention of Marine Pollution by Dumping of Wastes and Other Matter, 1972, 1975 • Memorandum of Understanding concerning Conservation Measures for Marine Turtles of the Atlantic Coast of Africa, 1999, 2008 • Memorandum of Understanding on the Conservation and Management of Marine Turtles and their Habitats of the Indian Ocean and South-East Asia, 2001 • Memorandum of Understanding on the Conservation of Migratory Sharks, 2010 • Montreal Protocol 1987 • Nairobi Convention, 1985, 1996 • Nairobi International Convention on the Removal of Wrecks, 2007, 2015 • Treaty on the Non-Proliferation of Nuclear Weapons, 1968, 1970 • Convention on Nuclear Safety, 1994, 1996 • Nuclear Terrorism Convention, 2005, 2007 • Ottawa Treaty, 1997[72] • Outer Space Treaty, 1967 • Paris Agreement, 2016 • Partial Nuclear Test Ban Treaty, 1963

72 Also relevant for Governance.

Environmental (South Africa)
• Protocol on Environmental Protection to the Antarctic Treaty, 1991, 1998 • Ramsar Convention, 1971, 1975 • Rotterdam Convention, 1998 • Seabed Arms Control Treaty, 1971 • Section 123 Agreement, 1954 • Statutes of the International Centre for Genetic Engineering and Biotechnology, 1983, 1994 • Statute of the International Renewable Energy Agency, 2009, 2010 • United Nations Convention on the Law of the Sea, 1982, 1994 • United Nations Convention to Combat Desertification, 1994, 1996
Social (South Africa)
• Abolition of Forced Labour Convention, 1957, 1959 • African Charter on Human and Peoples' Rights, 1981, 1986, 2004 • African Charter on the Rights and Welfare of the Child, 1990, 1999 • African Union Convention on Preventing and Combating Corruption, 2003 • African Youth Charter, 2006, 2009 • Convention on the Rights of the Child, 1989, 1990 • Committee Against Torture, 1987 • Constitution of the International Organization for Migration, 1951 • Convention Concerning Statistics of Wages and Hours of Work, 1938 • Convention on Domestic Workers, 2011, 2013 • Convention on Mutual Administrative Assistance in Tax Matters, 1988, 1995

(continued)

Table 6.12 (continued)

Social (South Africa)
• Convention on Psychotropic Substances, 1971, 1976 • Convention on Consent to Marriage, Minimum Age for Marriage and Registration of Marriages, 1962, 1964 • Convention on the Nationality of Married Women, 1957, 1958 • Convention on the Protection and Promotion of the Diversity of Cultural Expressions, 2005, 2007 • Convention on the Protection of the Underwater Cultural Heritage, 2001, 2009 • Convention on the Rights of Persons with Disabilities, 2007, 2008 • Convention Relating to the Status of Refugees, 1951, 1954 • Discrimination (Employment and Occupation) Convention, 1958, 1960 • Doha Amendment to the Kyoto Protocol, 1997, 2005 • Equal Remuneration Convention, 1951, 1953 • Final Articles Revision Convention, 1961 • First Optional Protocol to the International Covenant on Civil and Political Rights, 1966, 1976 Convention Against Discrimination in Education, 1960, 1962 • Freedom of Association and Protection of the Right to Organise Convention, 1948, 1950 • Geneva Declaration on Armed Violence and Development, 2006[73] • Genocide Convention, 1948

73 An over-aching treaty at the core of economic development in Africa and that should also be a core objective of ESG in Africa. For detailed information on this convention, signatories, the incidence of global violence and work done in relation to the convention after its adoption in 2006, the interested reader may visit http://www.genevadeclaration.org/.

Social (South Africa)
• Hague Adoption Convention, 1993 • Hague Convention for the Protection of Cultural Property in the Event of Armed Conflict, 1954, 1956 • Hague Protocol, 1955, 1963 • International Convention Against Doping in Sport, 2005, 2007 • International Convention on Maritime Search and Rescue, 1979, 1985 • International Convention on the Elimination of All Forms of Racial Discrimination, 1965, 1969 • International Covenant on Civil and Political Rights, 1966, 1976 • International Covenant on Economic, Social and Cultural Rights, 1966, 1976 • Kyoto Protocol, 1997 • Labour Inspection Convention, 1947 • Lusaka Accords, 1984 • Maputo Protocol, 2005[74] • Maritime Labour Convention, 2006, 2013 • Minimum Age Convention, 1973 • Montreal Convention, 1999 • Nairobi Convention, 1985[75] • Occupational Safety and Health Convention, 1981

(continued)

74 The first continental protocol on women's rights in Africa, which serves as instrument of the African Charter on Human and Peoples' Rights (also known as the Banjul Charter).

75 Also relevant to the Environment and Governance.

Table 6.12 (continued)

Social (South Africa)
• OECD Anti-Bribery Convention, 1997, 1999 • Optional Protocol on the Involvement of Children in Armed Conflict, 2000, 2002 • Optional Protocol on the Sale of Children, Child Prostitution and Child Pornography, 2000, 2002 • Optional Protocol to the Convention on the Rights of Persons with Disabilities, 2007, 2008 • Ottawa Treaty, 1997 • Protocol amending the Single Convention on Narcotic Drugs, 1972 • Protocol Relating to the Status of Refugees, 1967 • Protocol to the African Charter on Human and Peoples' Rights on the establishment of an African Court on Human and Peoples' Rights, 1998, 2004 • Protocol to the Convention against Discrimination in Education, 1960, 1962 • Ramsar Convention, 1971, 1975

Social (South Africa)
• Right to Organise and Collective Bargaining Convention, 1949 • Rome Statute, 1998 • Safety and Health in Mines Convention, 1995 • Second Optional Protocol to the International Covenant on Civil and Political Rights, 1989, 1991 • Single Convention on Narcotic Drugs, 1961, 1964, 1972, 1975 • Statute of the International Atomic Energy Agency, 1957 • United Nations Convention Against Illicit Traffic in Narcotic Drugs and Psychotropic Substances, 1988, 1990
Governance
• Abuja Treaty, 1991 • African Free Trade Zone, 2008 • Agreement on Technical Barriers to Trade, 1995 • Agreement on Trade-Related Investment Measures, 1994, 1995 • Antarctic Treaty System, 1959, 1961

(continued)

Table 6.12 (continued)

Governance
• Arms Trade Treaty, 2013, 2014 • Brazzaville Protocol, 1988 • Charter of the Indian Ocean Rim Association for Regional Co-operation, 1995 • Convention on Certain Conventional Weapons, 1981, 1983 • CITES, 1973, 1975 • Constitutive Act of the African Union, 2000, 2001 • Convention for the Mutual Recognition of Inspections in respect of the Manufacture of Pharmaceutical Products, 1971 • Convention for the Suppression of Unlawful Acts against the Safety of Civil Aviation, 1971, 1973 • Convention on the International Maritime Organization, 1948 • Convention on the International Mobile Satellite Organization, 1976, 1979

Governance
• Convention respecting the Laws and Customs of War on Land, 1899, 1907 • Cotonou Agreement, 2000, 2003, 2005, 2010 • Convention establishing a Customs Co-operation Council, 1953 • Final Articles Revision Convention, 1961 • General Agreement on Trade in Services, 1995 • Geneva Declaration on Armed Violence and Development, 2006 • Genocide Convention, 1948, 1951 • Hague Conventions of 1899 and 1907, 1988, 1907 • Hague Hijacking Convention, 1970, 1971 • Hague Protocol, 1955, 1963 • Hostages Convention, 1979, 1983 • International Convention on Maritime Search and Rescue, 1985, 1979, 1998, 2004 • Maputo Protocol, 2003 • Nairobi Convention, 1985 • Ottawa Treaty, 1997 • Outer Space Treaty, 1967 • Protocol I, 1977 • Protocol II, 1977 • Protocol on Incendiary Weapons, 1980, 1981, 1983 • Protocol on Mines, Booby-Traps and Other Devices, 1983, 1996, 1998 • Rome Statute, 1998 • SADC Treaty, 1980, 1992 • Settlement Proposal, 1978 • UNIDROIT Statute, 1926, 1940 • United Nations Convention on the Law of the Sea, 1982, 1994

Source Various—Wikipedia, authors, etc.

CHAPTER 7

An Empirical Analysis of ESG and Corporate Performance in South Africa

Abstract An empirical analysis of the relationship between ESG and corporate performance is carried out in this chapter, using South Africa as a case study. We cover the period from 2008 to 2018, using data for a total of 140 companies on the Johannesburg Stock Exchange (JSE). Our findings reveal that ESG disclosures are positively associated with financial and environmental performances and corporate governance is also positively related to financial performance.

Keywords South Africa · Corporate performance · CO_2 emissions · Sustainability

7.1 Introduction

Rigorous literature on what we refer to as formal or comprehensive ESG in this book is hardly existent on Africa. Even outside the continent research is nascent on ESG disclosure and corporate or firm performance. According to Friede et al. (2015), there is still a fragmentation of knowledge regarding the relationships. In this chapter we carry an original, empirical and more detailed study of the relationships between ESG and a number of selected parametres of corporate or firm performance in a

K. N. M. Sonko and M. Sonko, *Demystifying Environmental, Social and Governance (ESG)*, Palgrave Studies in Impact Finance,
https://doi.org/10.1007/978-3-031-35867-8_7

sample of listed South African companies on the Johannesburg Stock Exchange (JSE). Our study in this chapter is instrumental to further the quantitative understanding of the relationships between ESG and corporate performance, using South Africa as an example.

South Africa is an interesting and worthwhile case study on its own for a number of reasons. First, it is the second biggest economy in Africa and the most advanced in the continent by several measures of economic development, such as corporate governance, green financing, infrastructure, financial markets, and manufacturing (refer to Chapter 6). Politically, it is a very new democracy, grappling with conflicting interest groups as reflected in the huge number of parties, non-governmental organizations, and other evidence.[1] Racially and ethnically, it is still a divided nation, although apartheid has been eliminated. The economic potential is huge and some significant strides have been made by the government since 1994; nevertheless, it is struggling and, without the appropriate measures and the full support of the private sector, it risks running out of steam. The historical divide is gone officially, but it remains in reality to a very obvious extent. Violent riots in July 2021, following the trial of Zuma, clearly showed the immense socio-political issues the country faces, despite the immense opportunities.

Furthermore, the country is the only African member of the BRICS (Brazil, Russia, India, China, and South Africa) and the 20-member G20 (Group of 20). Moreover, it has traditionally attracted the highest levels of Foreign Direct Investments (FDI) over the years, in spite of the challenges the country faces.[2] Furthermore, the JSE is the largest, most developed, and sophisticated market in the Sub-Saharan African region (SSA) and is ranked among the top 20 stock markets worldwide (Nyasha et al., 2015) (Chapter 6).

South Africa, as reported earlier in Chapter 6, is a global pioneer of corporate governance codes. A good number of provisions of the

[1] With regard to political parties, the Electoral Commission of South Africa (IEC) announced on 20 March 2019 48 parties had registered to contest the national parliamentary elections. According to Wikipedia, the total number of parties (including those without a seat in parliament) is about 100, excluding defunct names. https://en.wikipedia.org/wiki/List_of_political_parties_in_South_Africa.

[2] For example, FDI inflows to the whole of Africa reached $83 billion in 2021, more than half of which ($42 billion) went to Southern Africa, almost all to South Africa with $41 billion, putting the country among the top 8 destinations in the world. See UNCTAD, *World Investment Report 2022*, UNCTAD, Geneva, 2022.

Corporate Governance Codes—the King Codes/Reports can be considered as shareholder oriented. Nonetheless, the reports also feature many stakeholder provisions, affirmative rules, initiatives, and actions. Since their inception, the existence of the Codes has persuaded and supported companies to include and disclose sustainability issues into their operations according to the Global Reporting Initiative (GRI) guidelines. The topics of Corporate Social Responsibility (CSR) and sustainability are well covered in the Codes, including the first and the subsequent Codes.

As further noted in Chapter 6, the JSE Limited is the second oldest, after the Egyptian exchange, and the largest exchange in Africa. It was founded on 8 November, 1887, and had market capitalization of US$ 1005 billion in 2020. The Exchange was established during South Africa's first gold rush, a commodity that was and continues to be central in the South African economy.

With regard to responsible investment, the JSE Socially Responsible Investment (SRI) Index was launched in 2004 and was replaced by the FTSE/JSE Responsible Investment Index Series. The JSE was the first stock exchange in the emerging markets to initiate/develop a sustainability index (Sonnenberg and Hamann, 2006). The particularity of this index is that it was adapted to the South African context, hence not just imitated in but harmonized to fit the South African society and business sphere.

In addition to all of the above reasons, the availability of market data makes South Africa the necessitous choice for the detailed quantitative case study in this chapter.

7.2 The Literature Review and Research Issues

The existing literature is based upon various theories. First, shareholder theory is the view that the unique and sole objective of an enterprise is to maximize its profit, thus financial performance for the benefit of shareholders. As in the case of the pre-ESG U.S. Business Roundtable cited in Chapter 1, South African corporations used to primarily follow this view, especially in the days of apartheid. They believed generally that the sole reason management works or must work on behalf of shareholders is to safeguard their interests and create maximum returns by way of increased share price or dividends (Friedman, 1970).

Dhaliwal et al. (2012) classified South Africa under a "shareholder-oriented economy" and only assessed the existence (not quality) of South

African Corporate Social Responsibility (CSR) reports. Hence, it is not clear if there is a justifiable link between sustainable performance and firm value. Also, it is important to point out that the research was conducted when the economy was still primarily shareholder-oriented. Since then, the country has initiated multiple efforts through the South African King Reports to evolve towards the direction of a relatively stakeholder-oriented economy A majority of criticism is that shareholders are only residual risk-bearers, hence the interests of all the different stakeholder groups and constituencies should be safeguarded (Flower, 2015; Schwab, 2021).

Stakeholder theory stipulates that a firm's sole objective is not to only maximize shareholders' interests. The interests of the various stakeholders (consumers, creditors, customers, employees, communities, etc.) should also be taken into consideration (Freeman & McVea, 2001). Hence, the idea of sustainable business and ESG are promoted, as this view supports the reduction of externalities and the maximization of social value.

Legitimacy theory is defined as a distinguished and continuous contract between society and companies. The latter choose to take up and endorse practices that are socially oriented to garner social approval (Guthrie and Parker, 1989). This social contract must be perpetually renewed and it legitimizes the right of existence of a firm. Hence, this theory puts forward as a basis for argument that ESG practices should manifest the intention of companies to embody a moral claim to an agreement reached with society (Scherer and Palazzo, 2007, 2011).

In this chapter, we conduct a study over a period of ten years (2008–2018) for a total of 140 listed companies (equivalent to about half of JSE's total number of listed companies (equity issuers) in 2023)[3] to enhance the calibre/quality of our results and thoroughly investigate any potential correlations. Therefore, we expand the existing literature on the subject through an increase of observations, two hypotheses, and relevant regression models. In order to address apposite issues and the gaps in the literature, we formulate the subsequent hypotheses.

[3] The JSE had 294 companies on the 15 January 2023, https://www.jse.co.za/.

7.3 The Hypotheses Developed

The literature review has provided multiple possible research questions and issues. There are a significant number of gaps, inconclusive and fragmented results in the literature. In this chapter we examine, specifically, the relationships hypothesized as follows:

Hypothesis 1 ESG disclosure is positively associated with financial performance (H1)

Hypothesis 2 ESG performance is positively associated with firm value (H2)

We use a long and relatively recent time period (2008–2018), utilizing ASSET4/Worldscope (DataStream) for ESG and financial related data, instead of KLD or CDP datasets. We complement policies and industry practices by adopting similar firm-value methodology (Tobin's Q and ROA) and control and board variables (board diversity, slack, new technology, leverage, etc.).[4]

7.4 Empirical Models and Variables

We use multivariate analysis to inspect the effects of the variables of ESG performance, firm performance, and ESG disclosure. The relationships between these three variables are represented through **H1** and **H2**. This is analysed by means of correlations and regressions in the fixed effect estimation method. Minutolo et al. (2018) also employed the fixed effect model (FEM) in order to study ESG disclosure effects on S&P 500 financial performance. We conduct the Hausman test, suggesting that once

[4] This chapter is part of a much broader study, in which we assess sustainability variables (GRI-guidelines) and time dummy variables (2012/2016). We categorize the firms according to industry practices, by singularly dividing the sample into polluting (85) and non-polluting (55) companies, thus totaling 140 firms, 1140 observations, for 10 years. Thereafter, the fixed model effect was utilized in order to analyse the data. The interested reader may refer to Mariama Sonko, "The Effects of Environmental, Social and Governance (ESG) Disclosure and Performance on Firm Value: Evidence from South African Listed Firms," dissertation submitted in partial fulfilment of the requirements for the degree of MSc. at Heriot-Watt University, Dubai Campus, 21/8/2019. In this study, ESG disclosures are measured individually (E, S, G) and in sum. This was done to achieve an adequate comparison of ESG disclosures measured separately and in sum over designated periods.

the p value is significant one may proceed to using the fixed effect model for the panel dataset. It is argued by Hsiao (2007) that the model of panel data offers better efficiency and adds accuracy conjectures because it controls omitted variable issues (unobservable or missing) and captures the heterogeneous "unobservables" amid individual units or gradually over the time period. Therefore, the FEM is fitting as it offers increased consistency and fewer bias results (Gallego-Alvarez et al., 2015).

Using two dependent variables (Tobin's Q and ROA), we develop the empirical models that follow.

Hypothesis 1 ESG disclosure is positively associated with financial performance (H1)

$$
\begin{aligned}
(1)\ TobinQ_{i,t} = {} & \beta_0 + \beta_1 ESG\_Disc\_Total_{i,t} + \beta_2 Board\_Diversity_{i,t} \\
& + \beta_3 Board\_Member\_Affiliation_{i,t} + \beta_4 Board\_Size_{i,t} \\
& + \beta_5 Total\_Stock\_Based\_Compensation_{i,t} + \beta_6 ISO\_14000_{i,t} \\
& + \beta_7 GRI\_Guidelines_{i,t} + \beta_8 Global\_Compact_{i,t} \\
& + \beta_9 Market\_Capitalisation_{i,t} + \beta_{10} Current\_Ratio_{i,t} \\
& + \beta_{11} Total\_debt\_to\_assets_{i,t} + \beta_{12} Mrkt\_value\_to\_book_{i,t} \\
& + \beta_{13} Employees_{i,t} + \beta_{14} Slack_{i,t} \\
& + \beta_{15} Ppe\_net_{i,t} + \beta_{16} Shareholders_{i,t} + u_{i,t}
\end{aligned}
$$

$$
\begin{aligned}
(2) ROA_{i,t} = {} & \beta_0 + \beta_1 ESG\_Disc\_Total_{i,t} + \beta_2 Board\_Diversity_{i,t} \\
& + \beta_3 Board\_Member\_Affiliation_{i,t} + \beta_4 Board\_Size_{i,t} \\
& + \beta_5 Total\_Stock\_Based\_Compensation_{i,t} + \beta_6 ISO\_14000_{i,t} \\
& + \beta_7 GRI\_Guidelines_{i,t} + \beta_8 Global\_Compact_{i,t} \\
& + \beta_9 Market\_Capitalisation_{i,t} + \beta_{10} Current\_Ratio_{i,t} \\
& + \beta_{11} Total\_debt\_to\_assets_{i,t} + \beta_{12} Mrkt\_value\_to\_book_{i,t} \\
& + \beta_{13} Employees_{i,t} + \beta_{14} Slack_{i,t} + \beta_{15} Ppe\_net_{i,t} \\
& + \beta_{16} Shareholders_{i,t} + u_{i,t}
\end{aligned}
$$

Hypothesis 2 ESG performance is positively associated with firm value (H2)

$$(1) TobinQ_{i,t} = \beta_0 + \beta_1 ESG\_Disc\_Total_{i,t} + \beta_2 Board\_Diversity_{i,t} + \beta_3 Board\_Member\_Affiliation_{i,t} + \beta_4 Board\_Size_{i,t} + \beta_5 Total\_Stock\_Based\_Compensation_{i,t} + \beta_6 ISO\_14000_{i,t} + \beta_7 GRI\_Guidelines_{i,t} + \beta_8 Global\_Compact_{i,t} + \beta_9 CO_2\_Emission_{i,t} + \beta_{10} Market\_Capitalisation_{i,t} + \beta_{11} Current\_Ratio_{i,t} + \beta_{12} Total\_debt\_to\_assets_{i,t} + \beta_{13} Mrkt\_value\_to\_book_{i,t} + \beta_{14} Employees_{i,t} + \beta_{15} Slack_{i,t} + \beta_{16} Ppe\_net_{i,t} + \beta_{17} Shareholders_{i,t} + u_{i,t}$$

$$(2) ROA_{i,t} = \beta_0 + \beta_1 ESG\_Disc\_Total_{i,t} + \beta_2 Board\_Diversity_{i,t} + \beta_3 Board\_Member\_Affiliation_{i,t} + \beta_4 Board\_Size_{i,t} + \beta_5 Total\_Stock\_Based\_Compensation_{i,t} + \beta_6 ISO\_14000_{i,t} + \beta_7 GRI\_Guidelines_{i,t} + \beta_8 Global\_Compact_{i,t} + \beta_9 CO_2\_Emission_{i,t} + \beta_{10} Market\_Capitalisation_{i,t} + \beta_{11} Current\_Ratio_{i,t} + \beta_{12} Total\_debt\_to\_assets_{i,t} + \beta_{13} Mrkt\_value\_to\_book_{i,t} + \beta_{14} Employees_{i,t} + \beta_{15} Slack_{i,t} + \beta_{16} Ppe\_net_{i,t} + \beta_{17} Shareholders_{i,t} + u_{i,t}$$

The variables in the multiple regression models are defined as follows:

A. Dependent Variables

First, for **H1** we examine the effect of ESG disclosure on firm value and in **H2** we examine the effect of ESG performance on firm value, following the models of Yang and Bassandori (2017) and Minutolo et al. (2018). Hence, a firm's financial performance is predominantly defined by two main indicators: Tobin's Q and Return on Assets (ROA). The latter stipulates how efficiently the company has used its assets in order to generate profit. This indicator measures profitability and is one of the most commonly used to determine corporate financial performance (Guenster, Bauer, Derwall and Koedijk, 2011; Roberts and Dowling, 2002; Tang and Bassandorj, 2017):

$$ROA_{it} = (Net\ Profit_{it})/(Total\ Asset_{it})$$

The former (Tobin's Q) is also widely used as it is a complete financial indicator describing a firm's performance. In this vein, Tobin's Q is employed within multiple studies to distinguish the management of a firm (Alexander and Bucholz, 1978; Ding, Ferreira and Wongchoti, 2016; Hillman and Keim, 2001; Seo, Moon and Lee, 2015; Yang and Baasandorj, 2017). As this indicator's calculation is quite complex, we follow Chung and Pruitt's (1994) approximation. Hence, it is a ratio of the market value of the company over its replacement cost. The ratio suggests that if the company is managing its resources and capabilities well, then there is added value and the assets' value is superior to their replacement cost. The following formula was then inserted into DataStream:

$$Tobin's\ q_{it} = (Market\ Capit + Pref\,Stock_{it} + Debt_{it})/(Total\ Asset_{it})$$

As per the methodology of Yang and Baasandorj (2017) we identify two models, one for each of the financial performance variables. Hypotheses 1 and 2 present Tobin's Q and ROA as dependent variables, whereas the explanatory variables include ESG disclosures, board diversity, GRI Reporting Guidelines, and control variables, among others, to measure the effects. Each of the models is analysed for the total number of companies ($n = 1140$).

Before acknowledging empirical results, we deduce that both indicators are expected to be positively associated with ESG disclosures and performance.

B. Independent Variables

We use total ESG disclosure scores, board size, total stock-based compensation, board diversity, board member affiliations, ISO 14000, GRI reporting guidelines, and UN global compact as independent variables. This is in order to study their impacts on the two dependent variables (Tobin's Q and ROA).

First, in order to measure the effects on firm value (Tobin's Q and ROA), we assess total ESG disclosure scores.[5] This indicator illustrates how firms perform in these three areas and they assess social responsibility (Bassen and Kovacs, 2008; Kocmanova and Simberova, 2012). However,

[5] For the more extensive study with five equations and a more elaborate methodology, the interested reader may refer to Mariama Sonko, ibid.

very common drawbacks when using this indicator are that components for each industry differ. Therefore, firms publish data according to various approaches due to the lack of commonly accepted and universal reporting standards and structures by firms (Eccle et al., 2012). Nonetheless, to remedy these drawbacks, the ASSET4 ESG database was used as it is a single framework with comparable standards/benchmarks.

We use a dummy variable that indicates a company's adoption of ESG-based compensation policy. The value of 1 (yes) is characterized by the adoption of a specific disclosure and the value of 0 illustrates a non-adoption of a specific disclosure (no) for each of the three areas. In the case of a controversy, the roles are reversed. Thereafter, an overall weighted average score was calculated (Total_ESG_disclosures).

Acknowledging the positive correlations we find, ESG disclosure is expected to be positively correlated with firm value. Separately, the various ESG disclosure dummy variables are calculated as follows in Table 7.1:

Secondly, the board-related test variables are measured by board size, total stock-based compensation (Cordeiro and Sarkis, 2008), board diversity, and board member affiliations. These variables are chosen according to the availability of data and the previous studies of De Villers et al., 2011, Mallin and Michelon, 2011 and Liao et al., 2015. The natural logarithms of board size, stock compensation, and average corporate affiliations of board members are used. All board-related variables are expected to have a positive association with profitability (ROA) and Tobin's Q (according to RBV as explained).

Particularly, Prado-Lorenzo and Garcia-Sanchez (2010) suggested that large boards are commonly ineffective/inefficient because of the free rider issue and higher conflict during the process of making decisions. Hence, this may generate a poor response to ESG topics according to agency theory. In contrast, RBV argues that a large, experienced, skillful, and knowledgeable board may ameliorate environmental performance due to the provision of skilled guidance about environmental dilemmas and after-effects. Also, larger boards can facilitate the obtainment of crucial financial and technological capital (De Villers et al., 2011). In this view, board affiliation is also likely to have a positive effect on ESG valuation and firm value.

Thirdly, the sustainability-oriented institutional variables are measured by ISO 14000, GRI reporting guidelines, and global compact. In accordance with the agency theory, these variables are predicted to have a

Table 7.1 ESG disclosure dummy variables construction

	Disclosure dummy variables			
Variable	Environmental disclosure	Social disclosure	Governance disclosure	Total ESG disclosure
Acronym	Env_disc	Soc_disc	Gov_disc	Total_ESG_disc
Formula	*= Σ (Biodiversity Impact Reduction + Clean Technology + Climate Change Risks/Opportunities Policy Emissions Reduction + Emissions Trading + Environmental Partnerships + Environmental Products + Product Environmental Responsible Use + Policy Energy Efficiency + PolicyWater Efficiency + Sustainable Building Products + Environment Management Team + Environment Management Training + Green Buildings + Renewable Energy Use)*	*= Σ (Policy Bribery and Corruption + PolicyBusiness Ethics + PolicyFair Competition + Business Ethics + Corporate Responsibility Awards + Crisis Management Systems + Policy Diversity and Opportunity + Policy Employee Health& Safety + HIV-AIDS Program + Health & Safety Management Systems + Policy Child Labor + Policy Forced Labor + Policy Human Rights + Human Rights Suppliers + ISO 9000 + Internal Promotion + Management Training + Six Sigma and Quality Management Systems + Policy Career Development + Policy Skills Training + Fundamental Human Rights ILO or UN + Product Access Low Price)*	*= Σ (Training Costs Total + Trade Union Representation + Donations Total)*	*= Σ(Env_disc + gov_disc + soc_disc)*

Source Authors. The variables are further defined in an appendix in the more extensive research referred to earlier (Sonko, 2019)

positive association with firm value and ESG disclosures, whereas inverse relationship with CO_2 emission is expected.

C. Control Variables

To control for company-specific determinants of ESG valuations and firm value, we follow, among others, De Villers et al. (2011) as well as Luo et al., (2012) and used control variables emanating from a number of corporate governance indicators. Indeed, we include firm size ("firm_size", natural logarithm of total assets), current ratio (current ratio), leverage ("total_debt_to_assets" ratio), market to book value ("Market to book ratio"), employees (natural logarithm of the number of employees), slack ("cash", ratio of cash and equivalents to total assets), new technology ("Property, plant and equipment (PPE) net"), and number of shareholders (natural logarithm of the number of shareholders) as control variables. Firm size, risk (debt to assets), and industry are one of the most commonly employed control variables to study sustainable performance and financial performance (Waddock and Graves, 1997; Voguel, 2005; Derwall, 2007; Nelling and Webb, 2009).

First, firm size's effect on ESG disclosure is of particular relevance. Waddock and Graves (1997) showed that smaller companies do not prioritize Corporate Social Responsibility (CSR) as compared to their larger counterparts. To that extent, bigger companies seem to commonly deal with added social and stakeholder demands to promote environmental and governance initiatives and ensure that they are increasingly effective (Waddock and Graves, 1997; De Villers et al., 2011; Liao et al., 2015). Therefore, the authors suggested that company size has a positive correlation with environmental performance (De Villers et al., 2011). Nevertheless, larger companies have to preserve their economic situation regarding their sales, products, and efficiency, therefore causing higher CO_2 emissions if there is no adoption of advance technology and energy efficiency, implying large capital investments.

Similarly, companies with numerous employees tend to have high rates of CO_2 emissions. RBV suggests that companies that generate higher profits and possess the appropriate financial resources tend to divert slack towards environmental and carbon programmes (De Villers et al., 2011).

Consistent with prior empirical results, firm size is expected to be positively associated with Tobin's Q and ROA. However, the effect on ESG disclosure is generally inconclusive.

Secondly, the debt to assets ratio is a reflection of a firm's capital structure and default risk dimension. It is assumed to be a key factor of CFP. If this ratio persists around a reasonable number, the firm has a good CFP. This ratio is used to control for risk since a management's outlook towards risk directly affects corporate performance (higher investment opportunities, build or destroy markets, cost management regarding costs to control pollution and improve labour conditions, etc.). Therefore, this ratio is assumed to be commonly negatively associated with ROA and Tobin's Q (Xhao et al., 2018).

Indeed, companies with larger debt to assets ratios possess higher obligations to pay interest, and thus causing a decrease of free cash flows and financial slack, in turn reducing environment related efforts. Additionally, firms with high debt tend to be able to sway and persuade managers to adopt short-term views on corporate activities and investments. Thereby, this leads to a reduction in corporate commitment to environmental activities (Haque, 2017). Therefore, this ratio is assumed to be commonly negatively associated with ESG performance.

Thirdly, McWilliams and Siegel (2000) defend the necessity of incorporating R&D investments and new technology as a control variable, to verify whether results are robust. The authors show a positive relationship between R&D investments and financial performance. Companies with a high number of new technologies are likely to use clean and energy efficient technology, hence improving the use of efficient energy and reducing CO emissions (De Villers et al., 2011; Liao et al., 2015). Correspondingly, this ratio is expected to have a positive impact on Tobin's Q and ROA.

Fourthly, companies with high market to book ratios are expected to have higher opportunities to invest, thus leading to superior environmental performance which increases their long-term competitive advantage (De Villers et al., 2011). This variable is used to carry out a robustness check to validate and corroborate results.

$$Market - to - book = (Market\ value\ of\ assets)/(Book\ value\ of\ assets)$$

Finally, enterprises with high shareholder numbers are assumed to suffer from high scrutiny and inspection, therefore they are forced to maintain advanced environmental and social standards (De Villers et al., 2011). This variable is expected to be positively associated with ESG disclosures and firm value.

7.5 Descriptive Statistics

Table 7.2 provides a list and summary of all variables. Table 7.3 indicates descriptive statistics of all variables used in the regression model.[6] The table shows that the mean value of environmental disclosure is 6.93, social disclosure is 13.72, and governance 11.28. This indicates that most South African listed firms commonly employ social disclosure more, as compared to governance or environmental disclosure. Indeed, the country has the most advanced CSR outlook in Africa due to the domestic demands/ pressure for CSR, the prevalence of mining in the economy, and the direct influence of corporate governance on CSR (Baskin, 2006). Also, the SRI Index is shown to be the most significant in the emerging economies (Baskin, 2006). Around 32% of the sample companies pursue ESG disclosure. This indicates the reality that the concept of comprehensive ESG disclosure was still not very widespread and most companies did not adopt such during the period.[7]

Additionally, only 18.67% of the sample firms have either female or foreign culture representation on the board. Despite efforts through a section on "diversity in the boardroom" in the King IV code, the most relevant exclusions are due to high social tensions, especially racial and gender inequality. These issues are also global phenomena, although South Africa's case is particular and unique in several important aspects because of its history. According to Deloitte's global survey in 2018, around 19% of JSE-listed companies have women directors with 6.9% being board chairpersons. The country is fourth globally for the percentage of women board members, against a global average of 4%. Also, the average board size of the sample firm is about 3. This signifies that boards are commonly quite small in size. The mean value of stock compensation is 16.32 and board affiliation is 1.13, these are relatively low values as compared to those of developed nations. These facts reveal

[6] A very extensive list of definitions of all variables in seven pages is available as an appendix from the authors.

[7] More recent evidence on South Africa and the other countries can be found in Chapter 6.

Table 7.2 Summary of all variables

No.	*Variable name*	*Variable nature*	*Acronym*	*Measurement criteria*	*Source of data*
1	Total ESG disclosure	Independent Variable (H1)	ESG_Disc_Total	See Table	ASSET4 ESG DataStream
2	Tobin's Q	Dependent Variable (H1) (H2)	Tobin_s_q	$Tobin = (Market\ Cap_{it} + PrefStock_{it} + Debt_{it})/(Total\ Asset_{it})$	WorldScope DataStream
3	Return On Assets	Dependent Variable (H1) (H2)	ROA	$ROA_{it} = (Net\ Profit_{it})/(Total\ Asset_{it})$	WorldScope DataStream
4	Board Diversity	Independent Variable (H1) (H2)	Board_Diversity	Natural logarithm of female and foreign culture representation on the board	ASSET4 ESG DataStream
5	Board affiliation	Independent Variable (H1) (H2)	Board_Member_ Affiliation	Natural logarithm of other corporate affiliations	ASSET4 ESG DataStream
6	Board Size	Independent Variable (H1) (H2)	Board_Size	Natural logarithm of board size	ASSET4 ESG DataStream
7	Total Stock-Based Compensation/Board compensation	Independent Variable (H1) (H2)	Total_Stock_ Compensation	Natural logarithm of stock compensation	WorldScope DataStream

No.	*Variable name*	*Variable nature*	*Acronym*	*Measurement criteria*	*Source of data*
8	ISO 14000	Independent Variable (H1) (H2)	ISO_14000	Dummy variable: 1 for the adoption, 0 for non-adoption	ASSET4 ESG DataStream
9	GRI Reporting Guidelines	Independent Variable (H1) (H2)	GRI_Reporting_Guidelines	Dummy variable: 1 for the adoption, 0 for non-adoption	ASSET4 ESG DataStream
10	Global Compact	Independent Variable (H1) (H2)	Global_Compact	Dummy variable: 1 for the adoption, 0 for non-adoption	ASSET4 ESG DataStream
11	Slack	Control variable (all hypotheses)	Cash	Ratio of cash and equivalents to total assets	WorldScope DataStream
12	Current Ratio	Control variable (all hypotheses)	Current_Ratio	Current assets/Current Liabilities × 100	WorldScope DataStream
13	Employees	Control variable (all hypotheses)	Employees	Natural logarithm of the number of employees	WorldScope DataStream
14	Firm Size	Control variable (all hypotheses)	Firm_Size	Natural logarithm of total assets	WorldScope DataStream
15	Market Value to Book	Control variable (all hypotheses)	Mrkt_value_to_book	Ratio of equity	WorldScope DataStream
16	Leverage	Control variable (all hypotheses)	Total_debt_to_assets	Total debt/total assets	WorldScope DataStream
17	Shareholders	Control variable (all hypotheses)	Shareholders	Natural logarithm of number of shareholders	WorldScope DataStream
18	New technology	Control variable (all hypotheses)	PPE_Net	PPE_net to gross PPE	WorldScope DataStream
19	Total CO_2 Emission	Control variable (all hypotheses)	CO_2_Emission-Total	Natural logarithm of total CO_2 Emissions	WorldScope DataStream

Table 7.3 Descriptive statistics

Variables	*Mean*	*Std. Dev*	*Min*	*Max*
ENV_DISC	6.93	2.69	0.00	13.00
SOC_DISC	13.72	4.21	0.00	19.00
GOV_DISC	11.28	8.14	0.00	21.00
ESG_DISC_TOTAL	31.93	11.44	0.00	46.00
BOARD_DIVERSITY	18.67	10.09	0.00	44.44
BOARD_MEMBER_AFFILIATION	1.13	0.50	0.22	2.57
BOARD_SIZE	2.50	0.22	1.79	3.14
TOTAL_STOCK_BASED_COMPENSATION	16.32	1.93	12.69	20.16
ISO_14000	0.77	0.49	0.00	2.00
GRI_REPORT_GUIDELINES	0.99	0.10	0.00	1.00
GLOBAL_COMPACT	0.42	0.50	0.00	1.00
TOBIN_S_Q	1.44	0.91	0.27	4.01
RETURN_ON_ASSETS	6.91	7.50	−14.59	24.77
CO_2 _EMISSION_TOTAL	13.62	1.07	11.30	15.67
FIRM_SIZE	16.56	1.25	13.52	19.60
MRKT_VALUE_TO_BOOK	3.02	2.85	0.22	12.46
PPE_NET	11.39	1.24	6.45	13.34
SHAREHOLDERS	15.72	2.05	7.84	18.02
TOTAL_DEBT___TOTAL_ASSET	15.59	10.72	0.00	45.41
CASH	7.08	1.47	2.50	10.90
CURRENT_RATIO	1.66	0.74	0.56	4.99
EMPLOYEES	9.22	0.94	6.97	11.47

Source Authors

a lot about the realities of South Africa in spite of the much-published subjects of black economic empowerment, gender equity, and other issues of equity addressed in the country's new (1994) constitution which is liberal by any standards.

The sustainability-oriented variables have high mean values, ranging from 0.42, 0.77 to 0.99. Hence, most companies have adopted either GRI, or global compact ISO 1400. Furthermore, the mean value of CO_2 emission is 13.62, although the level of carbon emissions is higher in the mining sector (highly represented in South Africa), succeeded by food and beverage and the industrial sector (Immink, 2018).

7.6 Correlations

Table 7.4 illustrates the bivariate correlations among all variables of the regressions.[8] On the one hand, it is noticed that there is a heteroscedasticity problem between market to book ratio and Tobin's Q. Also, this issue is noticed between CO_2 emissions and the number of shareholders. To make sure that these heteroscedasticity problems do not hinder the quality of our results, we make sure that the heteroscedasticity variables are not employed within the same regression models. Therefore, they did not pose an issue for the subsequent regression models which are formulated earlier; i.e.:

Hypothesis 1 ESG disclosure is positively associated with financial performance (H1)

$$
\begin{aligned}
(1) TobinQ_{i,t} = {} & \beta_0 + \beta_1 ESG\_Disc\_Total_{i,t} + \beta_2 Board\_Diversity_{i,t} \\
& + \beta_3 Board\_Member\_Affiliation_{i,t} + \beta_4 Board\_Size_{i,t} \\
& + \beta_5 Total\_Stock\_Based\_Compensation_{i,t} + \beta_6 ISO\_14000_{i,t} \\
& + \beta_7 GRI\_Guidelines_{i,t} + \beta_8 Global\_Compact_{i,t} \\
& + \beta_9 Market\_Capitalisation_{i,t} + \beta_{10} Current\_Ratio_{i,t} \\
& + \beta_{11} Total\_debt\_to\_assets_{i,t} + \beta_{12} Employees_{i,t} \\
& + \beta_{13} Slack_{i,t} + \beta_{14} Ppe\_net_{i,t} + \beta_{15} Shareholders_{i,t} + u_{i,t}
\end{aligned}
$$

$$
\begin{aligned}
(2) ROA_{i,t} = {} & \beta_0 + \beta_1 ESG\_Disc\_Total_{i,t} + \beta_2 Board\_Diversity_{i,t} \\
& + \beta_3 Board\_Member\_Affiliation_{i,t} + \beta_4 Board\_Size_{i,t} \\
& + \beta_5 Total\_Stock\_Based\_Compensation_{i,t} + \beta_6 ISO\_14000_{i,t} \\
& + \beta_7 GRI\_Guidelines_{i,t} + \beta_8 Global\_Compact_{i,t} \\
& + \beta_9 Market\_Capitalisation_{i,t} + \beta_{10} Current\_Ratio_{i,t} \\
& + \beta_{11} Total\_debt\_to\_assets_{i,t} + \beta_{12} Mrkt\_value\_to\_book_{i,t} \\
& + \beta_{13} Employees_{i,t} + \beta_{14} Slack_{i,t} + \beta_{15} Ppe\_net_{i,t} \\
& + \beta_{16} Shareholders_{i,t} + u_{i,t}
\end{aligned}
$$

[8] The correlation matrix is from the original study (Sonko, ibid.), which measures ESG disclosures separately and in sum, board variables, CO_2 emissions, control variables, and sustainability-oriented variables.

Table 7.4 Correlation matrix

Variables	*1*	*2*	*3*	*4*	*5*	*6*	*7*	*8*	*9*	*10*	*11*	*12*	*13*
ENV_DISC	1.00	0.53	0.16	0.55	0.13	0.01	0.28	0.22	0.22	0.15	−0.05	0.07	0.03
SOC_DISC	0.53	1.00	0.30	0.71	0.06	−0.09	0.06	0.16	0.07	0.18	0.19	0.14	−0.07
GOV_DISC	0.16	0.30	1.00	0.86	0.19	0.17	0.08	0.05	0.04	−0.21	0.06	0.14	−0.01
ENV_DISC_ TOTAL	0.55	0.71	0.86	1.00	0.19	0.09	0.15	0.15	0.10	−0.05	0.10	0.17	−0.03
TOBIN_S_Q	0.13*	0.06*	0.19*	0.19*	1.00	0.63	0.12	0.00	−0.16	−0.16	−0.30	−0.06	−0.06
RETURN_ON_ ASSETS	0.01	−0.09	0.17	0.09	0.63	1.00	−0.13	−0.25	−0.04	−0.21	−0.19	−0.06	0.00
BOARD_ DIVERSITY	0.28*	0.06*	0.08**	0.15**	0.12	−0.13	1.00	0.25	0.13	−0.02	−0.08	0.00	0.21
BOARD_ MEMBER_ AFFILIATION	0.22*	0.16*	0.05*	0.15**	0.00	−0.25	0.25	1.00	0.20	0.19	0.16	0.16	0.06
BOARD_SIZE	0.22*	0.07**	0.04*	0.10*	0.16	−0.04	0.13	0.20	1.00	0.45	0.29	0.05	0.06
TOTAL_STOCK_ COMPENSATION	0.15*	0.18**	−0.21*	−0.05*	−0.16	−0.21	−0.02	0.19	0.45	1.00	0.13	0.02	0.15
ISO_14000	−0.05	0.19	0.06	0.10	−0.30	−0.19	−0.08	0.16	0.29	0.13	1.00	0.16	−0.03
GRI_REPORT_ GUIDELINES	0.07	0.14	0.14	0.17	−0.06	−0.06	0.00	0.16	0.05	0.02	0.16	1.00	−0.12
GLOBAL_ COMPACT	0.03	−0.07	−0.01	−0.03	−0.06	0.00	0.21	0.06	0.06	0.15	−0.03	−0.12	1,00

***, **, and * indicate statistical significance at 1, 5, and 10% levels, respectively

Hypothesis 2 ESG performance is positively associated with firm value (H2)

$$
\begin{aligned}
(1) TobinQ_{i,t} = {} & \beta_0 + \beta_1 ESG\_Disc\_Total_{i,t} + \beta_2 Board\_Diversity_{i,t} \\
& + \beta_3 Board\_Member\_Affiliation_{i,t} + \beta_4 Board\_Size_{i,t} \\
& + \beta_5 Total\_Stock\_Based\_Compensation_{i,t} + \beta_6 ISO\_14000_{i,t} \\
& + \beta_7 GRI\_Guidelines_{i,t} + \beta_8 Global\_Compact_{i,t} \\
& + \beta_9 CO_2\_Emission_{i,t} + \beta_{10} Market\_Capitalisation_{i,t} \\
& + \beta_{11} Current\_Ratio_{i,t} + \beta_{12} Total\_debt\_to\_assets_{i,t} \\
& + \beta_{13} Employees_{i,t} + \beta_{14} Slack_{i,t} \\
& + \beta_{15} Ppe\_net_{i,t} + u_{i,t}
\end{aligned}
$$

$$
\begin{aligned}
(2) ROA_{i,t} = {} & \beta_0 + \beta_1 ESG\_Disc\_Total_{i,t} + \beta_2 Board\_Diversity_{i,t} \\
& + \beta_3 Board\_Member\_Affiliation_{i,t} + \beta_4 Board\_Size_{i,t} \\
& + \beta_5 Total\_Stock\_Based\_Compensation_{i,t} + \beta_6 ISO\_14000_{i,t} \\
& + \beta_7 GRI\_Guidelines_{i,t} + \beta_8 Global\_Compact_{i,t} \\
& + \beta_9 CO_2\_Emission_{i,t} + \beta_{10} Market\_Capitalisation_{i,t} \\
& + \beta_{11} Current\_Ratio_{i,t} \\
& + \beta_{12} Total\_debt\_to\_assets_{i,t} + \beta_{13} Mrkt\_value\_to\_book_{i,t} \\
& + \beta_{14} Employees_{i,t} + \beta_{15} Slack_{i,t} \\
& + \beta_{16} Ppe\_net_{i,t} + u_{i,t}
\end{aligned}
$$

Next, we analyse the correlations. First, **H1.I** is validated; ESG disclosure variables are positively correlated with Tobin's Q (as expected), statistically significant although weakly. In contrast, it is invalidated for ROA as none of the values are significant (**H1.II**).

Secondly, ESG disclosure variables are all positively correlated with three out of four board variables (board size, affiliation, and diversity). In contrast, governance disclosure and ESG total disclosures are negatively correlated with stock compensation. This would be music to those who in recent years have been complaining about high executive compensations. Also, board diversity and board affiliation are positively correlated with Tobin's Q, scores for governance "team". However, board compensation and size are weak and negatively correlated with Tobin's Q and all board variables are negatively correlated and non-significant with ROA.

Thirdly, we analyse the ESG performance indicator; CO_2 emissions. Fourthly, we analyse the sustainability-oriented variables. Total ESG disclosures variable is positively correlated with ISO 14000. Similarly, ESG disclosures are positively correlated with GRI Reporting Guidelines.

Lastly, we analyse the control variables. Every ESG disclosures variable is positively correlated with slack, firm size, and PPE net but negatively correlated with current ratio, as expected. Also, only environmental disclosures are negatively correlated with leverage. Solely social disclosures variable is negatively correlated with market value to book. Governance disclosures are negatively correlated with shareholders. Governance and total ESG disclosures are also negatively correlated with the number of employees. These are findings that make direct contributions to the literature and understanding of the main issues of and at the centre of ESG and the key debates about modern finance, management, and economic development.

Nonetheless, it is vital to further examine the multivariate regression results, prior to deducing statistical conclusions upon this matter. This is done in the next section.

7.7 The Results

This section describes fixed effect regression results.[9] Table 7.5 illustrates whether ESG disclosures affect Tobin's Q and *ROA* (**H1**). Table 7.6 illustrates whether ESG performance (measured by CO_2 emissions) affects Tobin's Q and ROA (**H2**). The findings are summarized in Table 7.7.

7.7.1 Tables and Regression Analysis

Table 7.5 shows the company fixed effect results of Tobin's Q (column 1) as a dependent variable and the company fixed effect results of ROA as a dependent variable (column 2).

Column 1 shows specification results of Tobin's Q against ESG disclosure total, board variables, and control variables. It is shown that the total ESG disclosure has a statistically significant (at 5% level) and weak positive association with Tobin's Q. *Numerically,* an increase by one

[9] For a more extensive study with five equations and a more elaborate methodology, the interested reader may refer to Sonko, ibid.

Table 7.5 Fixed effect regression results of Tobin's Q and ROA (Hypothesis 1)

Variables	*Tobin's Q*		*ROA*	
	1	1	2	2
Total ESG disclosure	0.010 (0.005)**	0.010 (0.005)**	0.040 (0.077)	0.072 (0.093)
Board variables				
Board diversity	−0.017 (0.007)**	−0.015 (0.007)**	0.177(0.116)	0.131 (0.138)
Board member affiliation	0.542 (0.174)***	0.460 (0.164)***	−4.720 (2.857)	−3.999 (3.301)
Board size	−0.724 (0.407)***	−0.748 (0.398)***	−10.777 (6.636)	−10.724 (7.965)
Total stock-based compensation	0.191 (0.057)***	0.105 (0.054)***	1.877 (0.939)***	2.168 (1.085)**
Sustainability-oriented institutional variables				
ISO 14000		0.259 (0.182)		−1.296 (3.702)
GRI		−0.080 (0.289)		−2.818 (6.405)
Global compact		−0.110 (0.197)		−0.208 (3.938)
Control variables				
Firm size	0.173 (0.080)**	0.165 (0.090)**	−0.484 (1.296)	−0.057 (1.678)
Current ratio	0.098 (0.081)	0.054 (0.084)***	3.839 (1.329)***	3.995 (1.691)**
Leverage	−0.011 (0.006)**	−0.012 (0.005)**	−0.169 (0.091)**	−0.197 (0.108)***
Mrkt value to book			1.862 (0.582)***	1.752 (0.554)***
Employees	−1.378 (1.663)	−1.117 (1.492)	−2.562 (26.925)	−2.742 (2.978)
Cash (Slack)	0.096 (0.098)	0.088 (0.094)	5.604 (1.597)***	5.455 (1.884)***
PPE Net	−0.060 (0.044)	−0.099 (0.053)***	−0.708 (0.722)	−1.463 (1.071)
Shareholders	−0.334 (0.130)**	−0.282 (0.126)**	−0.597 (2.138)	−0.685 (2.513)
Constant	14.557 (15.524)	13.680 (14.105)	21.635 (25.136)	24.066 (28.138)

(continued)

Table 7.5 (continued)

Variables	*Tobin's* Q		*ROA*	
VIF	1.039	1.037	1.374	1.325
Fstatistics	18.349***	15.476***	4.295***	3.159***
R-squared	0.836	0.833	0.547	0.347

Source Authors

***, **, and * indicate statistical significance at 1, 5, and 10% levels, respectively. The figures in parentheses are the standard errors

Table 7.6 Fixed effect regression results of Tobin's Q and ROA (Hypothesis 2)

Variables	*Tobin's Q*	*ROA*
	1	2
Board variables		
Board diversity	0.005 (0.005)	0.279 (0.180)
Board member affiliation	0.145 (0.097)	−4.299 (3.531)
Board size	−0.526 (0.247)**	−6.975 (8.959)
Total stock-based compensation	0.021 (0.031)	0.982 (1.129)
Sustainability-oriented institutional variables		
ISO 14000	0.114 (0.101)	−2.265 (3.680)
GRI	0.105 (0.164)	−2.925 (5.962)
Global compact	−0.057 (0.105)	0.118 (3.816)
CO_2 Emissions	−0.151 (0.065)**	−2.220 (2.377)
Control variables		
Firm size	0.119 (0.050)**	2.046 (1.826)
Current ratio	0.085 (0.047)***	2.145 (1.722)
Leverage	−0.011 (0.003)***	−0.177 (0.108)
Mrkt value to book		2.244 (0.690)***
Employees	−1.706 (2.852)	−6.608 (10.369)
Cash (Slack)	0.012 (0.054)	1.669 (1.952)
PPE Net	−0.027 (0.029)	−1.217 (1.043)
Constant	17.948 (26.183)	27.360 (31.066)
VIF	1.038	1.234
Fstatistics	17.081***	3.471***
R-squared	0.961	0.550

Source Authors

***, **, and * indicate statistical significance at 1, 5, and 10% levels, respectively. The figures in parentheses are the standard errors

unit in Tobin's Q implies an increase by 0.011 in total ESG disclosure. Among the control variables, firm size is weakly statistically significant and positively associated with Tobin's Q. Also, current ratio and slack are positively associated but not significant with Tobin's Q. In the same vein, employees are negatively correlated with the former. Contrarily, leverage and shareholders are weakly statistically significant and negatively associated.

Column 2 shows specification results of Tobin's Q against ESG disclosure total, board, sustainability-oriented institutional variables (SOIV), and control variables. It is evident that the explanatory powers of ESG and board variables remain unchanged. Once again, SOIV are not significant

Table 7.7 Summary of overall findings

Hypothesis	*Dependent variable*	*Validated or invalidated*	*Key findings*
H1: ESG disclosure is positively associated with financial performance	H1.1 Tobin's Q	Validated	Theoretically, legitimacy and stakeholder theory posit that ESG strengthens the social contract between the company and multiple stakeholders, leading to high operating performance (Derwall, 2007; Flammer, 2013)
	H2.2 ROA	Invalidated	Empirically, Venanzi (2012) found no significant relationship between sustainability disclosures and ROA Our results might be due to subjective quantitative measurement of ESG disclosures, lack of consensus, and widely accepted reporting standards with respect to ESG disclosure/performance in South Africa. Also, other financial indicators may be more significant such as ROI
H2: ESG performance is positively associated with financial performance	H2.1 Tobin's Q	Validated	ESG performance (as measured by $C0_2$ emissions) is negatively associated and statistically significant with tobin's Q
	H2.2 ROA	Invalidated	ESG performance (as measured by $C0_2$ emissions) is not statistically significant with ROA In conclusion, ESG performance increases operating capacity but does not impact profitability. This may signify that managers focus on short-term targets rather than long-term strides to reduce ESG performance (CO_2 emissions)

Source Authors

and remain around similar values. Moreover, firm size and current ratio have statistically significant and weak positive associations with Tobin's Q, whereas leverage, ppe, and shareholders have statistically significant and weak negative associations.

Columns 3 to 4 show that, in contrast to the results deduced for Tobin's Q, there is no significant effect of ESG disclosure variables on profitability. *Columns 3 and 4* illustrate specification results of ROA against ESG disclosure variables, board variables, and control variables. It is shown that board compensation is positively related and significant with ROA. Additionally, current ratio, market value to book, and cash are positively associated and statistically significant with ROA; whereas leverage is inversely related and significant. Notably, employees are negatively associated with ROA but not significant.

Table 7.6 shows fixed effect regression results of Tobin's Q and ROA on ESG Performance (measured by CO_2 Emissions). *Column 1* shows specification results of Tobin's Q against board, SOIV, and control variables and most importantly CO_2 emissions. The results show that board size is the only negative and statistically significant board variable. The rest are positively associated but not significant. Additionally, SOIV are not significant. ISO 14000 and GRI are positively associated with Tobin's Q, whereas global compact is negatively associated. Moreover, CO_2 emission is statistically significant and negatively associated with Tobin's Q. Lastly, among the control variables, firm size and current ratio have statistically significant and weak positive associations with Tobin's Q, whereas leverage has a statistically significant and weak negative association. *Column 2* shows specification results of ROA against board, SOIV, and control variables and also CO_2 emissions. We notice that CO_2 emissions are negatively associated with ROA but not statistically significant.

Overall, estimation results are mostly in agreement that total ESG disclosures and Tobin's Q are positively associated and Tobin's Q and CO_2 emission are positively associated. These disclosures are largely insignificant in relation to corporate profitability (ROA). Correspondingly, board variables are insignificant with total ESG disclosures.

7.7.2 Hypotheses Deductions

Hypothesis 1 ESG disclosure is positively associated with financial performance (Tobin's Q). The hypothesis H1.1 has been validated.

Theoretically, it is in line with legitimacy and stakeholder theories as ESG strengthens the social contract between the company and multiple stakeholders. Firm value also echoes investors' perceptions. High ESG disclosure translates to high market valuation, due to the perception of low risk and low discount rates (Derwall, 2007; Flammer, 2013). Empirically, these results are validated by previous studies examining corporate financial performance and corporate social performance and/ or ESG (Orlitzky et al., 2003; Margolis and Walsh, 2003; Minutolo et al. 2018).[10]

Hypothesis 2 ESG performance affects financial performance (Tobin's Q) but not ROA. Hypothesis H2.1 has been validated and hypothesis H2.2 has been invalidated.

CO_2 emissions (ESG performance) is negatively associated and statistically significant with Tobin's Q, whereas it is not statistically significant with ROA. Hence, ESG performance increases operating capacity but does not impact profitability, according to this study.

Empirically, Venanzi (2012) found no significant relationship between sustainability disclosures and ROA. Also, Barnett and Salomon (2012) found insignificant results when they used the Fixed Effects model. However, Almeyda[11] and Darmansyah (2019) conducted research on companies in the real estate sector from the G7 countries, using ROA, ROC, Stock Price, and P/E. The panel data over five years (2014–2018), using multivariate regressions indicate that there is a statistically significant positive relationship between ESG disclosure with ROA and ROC, but not significant with Stock Price and P/E.

Finally, amid control variables, firm size is commonly seen to have a positive and statistically significant correlation with dependent variables, which confirms that larger firms practice higher ESG disclosure and have higher firm values (as mentioned in Chapter 2). Leverage is commonly seen to have a negative and statistically significant correlation with dependent variables, which confirms that firms with higher debt practise less ESG disclosure and have lower firm values. Lastly, the employee variable is commonly seen to have a positive and statistically significant correlation

[10] A research report by the World Bank Group (WBG) and the Government Pension Investment Fund (GPIF) of Japan on ESG and fixed income funds specifically states that "evidence suggests that incorporating ESG.

into fixed income investing should be part of the overall credit risk analysis and should contribute to more stable financial returns. It also dispels the myth that incorporating ESG means having to sacrifice financial returns" (Georg Inderst and Fiona Stewart, *Incorporating Environmental, Social, and Governance (ESG) factors into Fixed Income Investment*. Washington, DC: GPIF & World Bank Group, 2018).

[11] Raisa Almeyda and Asep Darmansyah, "The Influence of Environmental, Social, and Governance (ESG) Disclosure on Firm Financial Performance," *IPTEK Journal of Proceedings*, Series No. 5 (2019): 278–290.

with ESG disclosure and CO_2 emissions variables, which confirms that firms with a high number of employees practise more ESG disclosures and CO_2 missions.

7.8 Conclusions

ESG disclosure is positively associated with financial performance (Tobin's Q) (*H1*). Theoretically, this is validated by legitimacy and stakeholder theory because ESG strengthens the social contract between the company and multiple stakeholders. High ESG disclosure translates to high market valuation, due to the perceptions of low risk and low discount rates. ROA does not influence ESG disclosure/performance.[12]

Overall, estimation results are mostly in agreement that: (i) total ESG disclosures and Tobin's Q are positively associated and; (ii) ESG performance and Tobin's Q are positively associated. These are very significant findings and they corroborate arguments and past evidence that ESG disclosures are good for overall financial performance. This means that they are good for the firms themselves, first and foremost, which is our position in this book as argued from the beginning in Chapter 1. This also means they are good for the economy and the wider society. Although these disclosures are largely insignificant or inconclusive in relation to corporate profitability (ROA), this does not refute arguments for the desirability of formal or comprehensive ESG reporting. Correspondingly, board variables are insignificant with total ESG disclosures.

Board diversity and board affiliation turned out to be positive and significant with financial performance. They are also inversely related with CO_2 emissions, when we do the independent sample t-tests.[13] This highlights the importance of corporate governance. The fact that we do not find any significant relationship between ESG disclosures/performance and ROA, again is proof that if ESG disclosures do not help the profitability of firms (ROA), they do not hinder it either.

Future empirical research may investigate the relationship between ESG and other financial indicators such as return on capital employed

[12] For case for why ESG leades to higher value for companies, see for instance, Witold Henisz, Tim Koller, and Robin Nuttall, "Five Ways That ESG Creates Value," *Mckinsey Quarterly*, November 2019.

[13] See study mentioned in footnotes 4, 5, 8, and 10.

(ROCE) and some of the other standard financial performance indicators the authors used in a study of conventional and Islamic banks in the Gambia.[14] They could also look closer at the relationships between gender, ethnic, and racial diversity and ESG disclosures. Moreover, cross-country comparisons between South Africa and other African markets in the years ahead should be worthwhile and interesting. Finally, according to data availability, both the number of sampled companies and years may be increased and updated.

[14] Sonko, Karamo NM, and Mariama Sonko, "Promoting Islamic Banking in Africa: Top Bottom or Bottom Up? A Case Study of Banking in the Gambia," in *Islamic Finance as a Complex System*, eds. Haider A. Khan, and Karamo NM Sonko. Lanham, MD: Lexington Books, 2020.

PART IV

The Way Forward

CHAPTER 8

ESG: The Way Forward for Stakeholders

Abstract In this chapter, we first summarize the seven earlier ones of this book. Then we make very important additions to the contents at the institutional/organizational and individual levels in Africa, whilst incorporating global implications too. Further contributions are made to the topic of ESG through discussions of the role of governments, Africa's ESG priorities, avoidance of copy and paste, decoupling, stakeholder engagement, mutual benefit, greenwashing, corporate governance, carbon capitalism, technology, and the common guilt. It is argued that the responsibility for a better world, particularly through better environmental management, is everyone's because the guilt is everyone's.

Keywords Stakeholder engagement · Governance · Women · Children · Youth · Climate

8.1 Summary

In providing an overview and addressing the relevance of ESG in Africa in Part I, this book started with an assessment of ESG, its importance, origins, and emergence. This was done in Chapter 1. We defined it, traced the evolution and its causes, identified the actors, found out that it is a

K. N. M. Sonko and M. Sonko, *Demystifying Environmental, Social and Governance (ESG)*, Palgrave Studies in Impact Finance,
https://doi.org/10.1007/978-3-031-35867-8_8

new area of research on a combination of very old and very new issues. We ended the chapter guiding the reader through the structure of the book.

In Chapter 2, we addressed the necessity and applicability of ESG in Africa. It was shown conclusively that the continent's environmental, social, and corporate governance challenges are such that a compelling and comprehensive approach to ESG is essential, requiring many players or contributors. Here, we went further to demonstrate the not-obvious relationship between the environment and equity.

ESG is an enabler of sustainability and as such it requires Strategic Management in order to implement it comprehensively and most efficiently, especially during and in the aftermath of a global pandemic or economic crisis. COVID-19 is just an example, unfortunately our world is rather sickly widespread diseases and epidemics have become rather common, itself a clarion call for ESG! Therefore, in Part II, Chapter 3, we came up with and presented an innovative Strategic Management framework, demonstrating the relevance and multipronged approach of the framework we referred to as the African PESTLE Analysis (APA). Through the APA we demonstrated at length how ESG goals can be pursued in targeted, organized, systematic actionable ways. We also discussed the challenges faced by the modern company, introduced the APA and the relationship with ESG, the conventional PESTLE Analysis and well-known SWOT Analysis, and showed how it is relevant to personal challenges such as suicide and divorce.

Focusing on a pandemic the depth and intensity of which the modern world has perhaps never seen before, Chapter 4 assesses the impact of the COVID-19 virus on Africa and the essence of ESG and Strategic Management through the APA. We argue that because of the COVID-19 pandemic, business as usual must not be conducted in Africa; COVID-19-induced challenges require conducting business in manners that are far beyond the normal concerns and methods of "business as usual". We illustrate the global economy before and after the pandemic, national and global anti-COVID-19 measures, Africa's "mystery" of COVID-19 numbers, what can be done for Africa's health sectors through ESG, the need for clarity of roles and effective coordination. We maintain a word of caution that the Pandemic might be heading for an end or even have ended, but the threat is not. This, in addition to the threats of others we bring to our attention in this book calls for a more caring world of common destinies such as ESG.

In Chapter 5, we extend the more practical aspects of the understanding and management of ESG by delving into the designing and reporting of ESG disclosures. Our target is the ESG practitioner or anyone who wants to have a broad understanding of the concepts. We outline what factors to take into consideration, the process of designing and reporting, opportunities that make the outputs more attractive, and give real-life examples including our own.

Part III of the book, we provide detailed regional case studies of ESG in all the four main regions of Africa—East, West, Southern, and North. In this chapter, we highlight the economies showing the biggest locally owned and foreign companies, relevant national laws and international agreements, the stock exchanges, and NGOs. We answer numerous questions about ESG practices and potential in Egypt, Kenya, Nigeria, and South Africa, shedding light on historical and contemporary aspects of ESG in the economies of each country. This is done in Chapter 6.

In Chapter 7, we conduct an original and empirical analysis of ESG and corporate performance in South Africa independently of the other case studies. The purpose is to examine this topical relationship between ESG and corporate performance in the country, specifically, and the implications for other countries, generally. Through a study of 140 listed companies on the JSE over a period of ten years, we found that ESG disclosure is positively associated with financial performance and ESG performance is positively associated with firm value.

Finally, in this last part (IV) and Chapter 8 of the book, we summarize and conclude, using anecdotal evidence and some stories to bid farewell.

8.2 Food for Further Thoughts and Emphases

8.2.1 Governments and ESG

First, among our concluding remarks, should be a discussion of the role of the government/state in ESG. As it is now, ESG is considered a responsibility of the private sector, in particular big companies making "big" money! Governments, governmental organizations, and NGOs (national and international) can be divided into the "dumb", the preachers, and the "pouncers". The "dumb" keep quiet, the preachers talk, and some of them support and encourage with their policies and the "pouncers" attack their targets as if they themselves are not part of the problem in some way or at some level. Both the preachers and the "pouncers" are a mixture those who are genuine and those who are not, some even agents or representatives of third parties as evidence has shown. Of these, our radar is on the governments at the moment, because we may point at anybody but in reality it is they who run or must run the world. They set up and own the armies, central banks, laws and courts, and the physical infrastructure the private sector cannot exist without. On top of that they collect taxes and/or other dues.

Consequently in the area of ESG, an informed, exemplary, and inspiring government can only encourage private companies in its territory through various means to have ESG programmes. The opposite genre of government should not be expected to do so. Our African governments then must move from spectators to active regulators in the case of obligatory environmental practices and more active inspirators in all voluntary ESG spheres. Companies can be inspired to undertake ESG pursuits through investment incentives such as tax concessions, training allowances, rural and employment benefits, and special recognitions for ESG leaders, not laggards.

8.2.2 Africa's ESG Priorities

Second is the importance of the prioritization of ESG issues in Africa depending on the needs of countries, environments, and communities. We believe that given the development and underdevelopment challenges of Africa, in short defeating pervasive poverty and achieve sustainable progress, the continent's ESG priorities should, arguably, be in the following order of greatest priorities when choices have to be made:

1. Environmental conversation, without which humankind cannot survive
2. Community development, which can be inclusive and beneficial to all those who are in need the most of both sexes, all ages, anywhere
3. Empowerment of women and protection of children
4. Engagement and employment of the youth
5. Corporate governance (Fig. 8.1).

All the above have been discussed in this book and we believe that they should not only be topics of intellectual debates of academics, policy

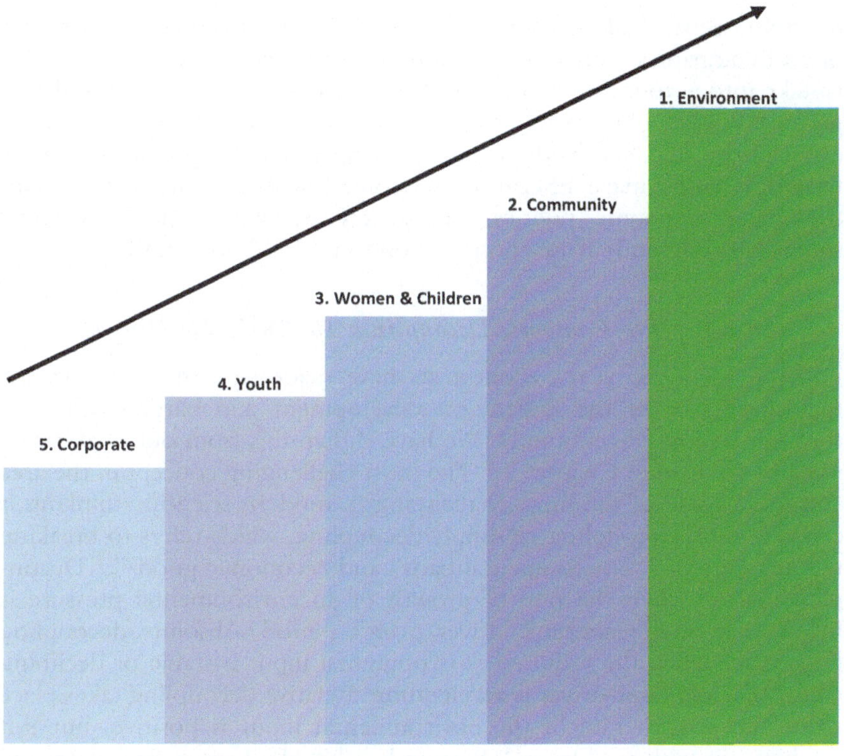

Fig. 8.1 Prioritization of Africa's ESG priorities (*Source* Authors)

makers, and politicians but seriously taken into account by all, especially in the ESG pursuits of private companies/businesses.

8.2.3 *Thou Shall not Simply Copy and Paste*

Thirdly, and related to the preceding point, is that there must be originality in adopting ESG practices. As Motlanthe (2022) emphasized: "it is important to recognize that each country is different and therefore has different needs, challenges and capabilities".[1] For these reasons, advocates of and companies with ESG programmes (especially foreign ones) need to take into account each country's culture, customs, economy, and religion and avoid ESG extremism or fundamentalism which set the same expectations and standards as the developed western countries. Some flexibility would make ESG more welcomed in African and other countries where economy, politics, and society are woven and intertwined somewhat differently from those in London, New York, and Paris.

8.2.4 Eco-Economic Decoupling: The Desirable Divorce

Fourthly, over the years, economists have made sure that they occupy their fair place in the debate on development and natural resources through ecological economics. We have shown this from our own earlier work in the field in Chapter 2.[2] The most challenging concept in the area and in fact one of the most challenging in modern scientific thinking is eco-economic decoupling or simply decoupling, which refers to breaking the link between "environmental bads" and "economic goods".[3] Decoupling occurs when the rate of growth of an environmental pressure is below that of its economic driver over a period. Absolute decoupling takes place when the required environmental input is stable or declining whilst the economic driver is accelerating. Relative decoupling takes place when the growth rate of the environmental input is positive, but not catching up with the driver.[4] It sounds a bit like having your cake and eating it, but with human ingenuity and commitment, not too many things are beyond the bounds of possibilities especially we think and try hard enough. The Organisation for Economic Co-operation and Development (OECD) announced the beginning of its journey on this road

[1] Kgalema Motlanthe, "Remarks at the First High-Level Webinar on ESG and Stakeholder Capitalism in Africa," Virtual Conference organized by Heeno International and partners on 28 January 2022.

[2] Referring to the first author's.

[3] OECD, "Indicators to Measure Decoupling of Environmental Pressure from Economic Growth," 2002, p. 1, https://www.oecd.org/environment/indicators-modelling-outlooks/1933638.pdf, retrieved March 28, 23.

[4] OECD, ibid; and https://www.oecd-ilibrary.org/sites/23508323-en/1/3/2/index.html?itemId=/content/publication/23508323-en&_csp_=d6be40bd4ae0c5f453d60dfd6e32b9dd&itemIGO=oecd&itemContentType=book.

by including decoupling environmental pressures from economic growth (i.e. GDP) as one of the main objectives in its Environmental Strategy for the First Decade of the 21st Century, adopted by OECD Environment Ministers in 2001.

The Organization's experience may be instructive. The reader may be surprised at how optimistic and positive the OECD's was. In its study of 31 decoupling indicators representing a range of environmental matters, it assessed the possibilities of decoupling of environmental pressures from total economic activity. 16 indicators covered natural resources, waste disposal, climate change, air pollution, water quality, and material use, whilst the rest of the indicators were spread over the production and usage in the energy, transport, agriculture, and manufacturing sectors. The Organization reported that relative decoupling is indeed common in the OECD. Absolute decoupling was also found to be widespread, except for some pressures, and was recorded in at least one country for all but two of the selected decoupling indicators at the country level. The OECD, therefore, concluded that "The evidence also suggests that further decoupling is possible".[5]

The OECD report sounds very simple on an extremely complicated matter. Not surprisingly it was met with outright rejection and a skew of criticism in some important quarters: "No evidence" (Vadén et al. (2020) found after a meta-analysis of 180 studies), and "nonsense" (according to Smil, 2019).[6]

We present the concept of decoupling in the context of the Environmental Kuznets Curve, which shows how, historically and generally, some decoupling tends to take place as economies grow and mature. Point A in Phase 1 of the shows an idealistic state which can only exist in what some may call "primitive" existence, but we would call simple, traditional, non-materialistic, or non-developmental societies, living mainly on need, in harmony with the environment. Point B suggests the idealistic opposite—the extreme form of an advanced modern service economy of

[5] OECD, ibid.

[6] Vaclav Smil, "'Growth must end. Our economist friends don't seem to realise that'", *The Guardian*, 21st September 2019. Retrieved March 28, 2021. The interested reader may also refer to the European Environmental Bureau (EEB), *Decoupling Debunked: Evidence and Arguments against Green Growth as a Sole Strategy for Sustainability*, 2019 (page visited retrieved 28th March, 2023).

the type OECD is referring to and which we call utopia. The inverted-U in the modified or stylized Environmental Kuznets Curve is only a depiction of hypothetical or "real" intertemporal relationships or changes. Much more work on the desirable potential "divorce" of economy and natural resource is needed from the business, social scientific and scientific communities, with the support of governments. This calls for the kind of coordination we recommend, explicitly and implicitly, in this book (see, for instance, Chapter 4).

No matter where you stand or sit in relation to the fence, the OECD study and declarations have generated a useful debate and drawn our attention to some of the key solutions to the painful marital problems of the eco-ecological conjugal union, apart from or in addition to divorce. These include the potential influences of structural and technological variables which indeed made differences in energy intensity and final consumption, GDP growth in the OECD. The experience also reflects the importance of effective governance in managing climate change through policies that are not only put on paper but implemented in reality resulting in energy operators reducing Sulphur oxide (SOx) emissions per unit of energy production (Fig. 8.2).[7]

8.2.5 Stakeholder Mapping and Engagement

Fifthly, at the micro or field level, stakeholder mapping and engagement are essential for every company.[8] A company must know who are its stakeholders, all those whom its needs to know and all it needs to know. It also needs stakeholder policies that are consistent and very good lines of communication and interaction with the stakeholders through the right community engagement or relations staff.

8.2.6 Good for the Goose and Gander

Sixth, there is an English idiom that "What is good for the goose is good for the gander"! The subject of ESG is such and it is being argued as such in this book. We maintain that formal comprehensive or holistic ESG is

[7] OECD, ibid.

[8] Check Matt's email, Perseus Sustainability Report and Chapter 5 (designing) on this point.

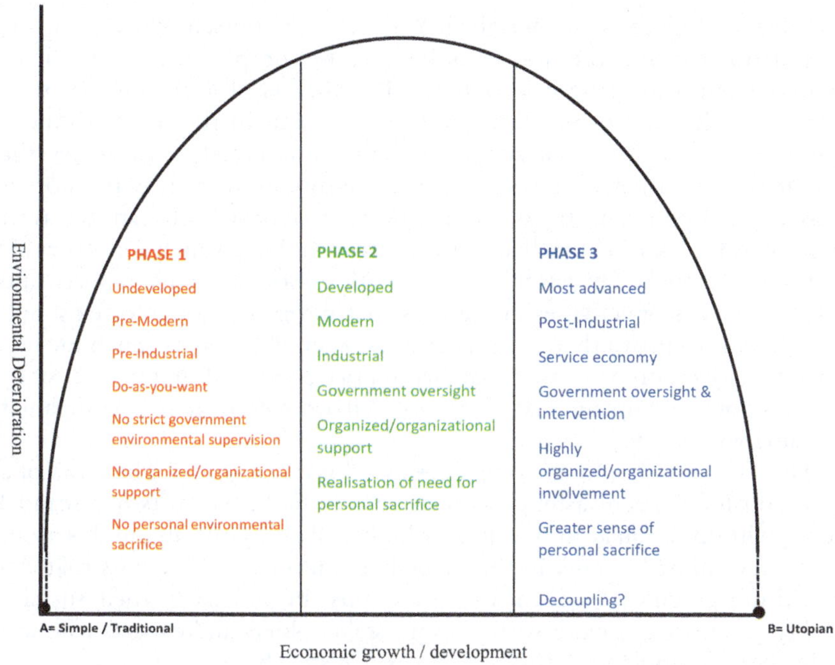

Fig. 8.2 The Environmental Kuznets curve with desirable divorce (*Source* Authors, inspired by Kuznets and the OECD)

good for businesses, economies and societies and nations and persons; in short, good for the world. Therefore, its cultivation is in a universal common where everyone has to till in his or her own interest and for the benefit of others. We can do so successfully through multiple players, on multiple fronts and all hands on deck.

8.2.7 Greenwashing

Seventh, as the name implies, greenwashing or climate washing is when a company portrays its activities or products to be more environmentally friendly than they actually are.[9] The purpose is to mislead—customers,

[9] For more readings: K. Becker-Olsen, and S. Potucek, "Greenwashing," In: *Encyclopedia of Corporate Social Responsibility*, ed. S. O. Idowu, N. Capaldi, L. Zu,

stakeholders, or the general public. You may ask yourself why a company would do that and the answer is because of the pressure for profit or public pressure to go green or both. The threats of legal suits for such actions in the case of tangible products are high in the most advanced countries where the legal system directly or indirectly encourages the public to sue for many things.[10] Even there though there may be another reason for a company to greenwash because it believes it can get away with it in the case of another form of greenwashing which can go easily unnoticed as such. The latter includes subtle announcements, reports, and advertisements of activities or features that never took place or exist and are difficult to prove the non-existence of, even if necessary. Such include an impressive company sustainability report produced at much cost by experts for a company that does not tell or wholly tells a company's sustainability story.

How or why should all companies avoid greenwashing? The costs of a legal suit for a greenwashing case can be very high, the costs of a scandal (even without a suit) can be equal or higher. Bearing this in mind, serious companies must be frank in their labelling and pronouncements that can constitute greenwashing. Instead, they must be honest in their sustainability practices and work with relevant stakeholders and experts who can help them to implement these practices realistically.

Even in Africa where the threats for lawsuits may not dissuade companies from greenwashing, honesty in marketing is what is desirable. The

and A. D. Gupta. Berlin, Heidelberg: Springer, 2013, https://doi.org/10.1007/978-3-642-28036-8_104; Miriam A. Cherry, "The Law and Economics of Corporate Social Responsibility and Greenwashing," *UC Davis Business Law Journal*, Vol. 14, No. 2: 281–303, https://scholarship.law.stjohns.edu/cgi/viewcontent.cgi?article=1498&context=faculty_publications; B. Elmore, "True Green or Pale Intention?" *Baylor Business Review*, Vol. 2 (2009): 48–49; N. E. Furlow, "Greenwashing in the New Millennium," *The Journal of Applied Business and Economics*, Vol. 10, No. 6 (2010): 22–25; E. Orange, and A. M. Cohen, "From Eco-Friendly to Eco-Intelligent," *The Futurist*, Vol. 44, No. 5 (2010): 28–32.

[10] In the European Union, for instance, numerous actions are being taken to tighten the screws against greenwashing. In March, 2023, alone, some of these include: (1) the European Parliament voted for a tougher law to protect consumers against commercial practices that are unfair, on March 28; (2) on the same day, they also banned all claims of reduced, neutral, positive, or compensated CO_2 impacton the basis of carbon offsetting; (3) the European Commission proposed the Green Claims Directive to address unreliable and misleading green marketing and establish standards for this, on March 22; Sources: https://eeb.org/, retrieved March 28, 2023.

sustainability practices of companies must be honestly reported annually or in any other intervals, as widely provided and explained in our menu of reporting forms or formats in Chapter 5. A rapidly increasing number of companies in and outside Africa are aligning their ESG programmes in theory or in practice to the UN SDGs, which are straightforward and universally accepted. National Development Plans (NDPs) in Africa are conceptually desirable too, but they can be controversial and plunge private businesses into political arenas they must not play football in.

8.2.8 Corporate Governance

Our eighth point is the fact that we sway a bit in this book between corporate governance and the environment in ESG; i.e. which should come first in ESG? It is not an easy decision! Good corporate governance is essential for effectively and efficiently achieving environmental, social, and financial goals of an enterprise. So, perhaps, ESG should be GES! However, the global climate crisis is so serious today that it is not easy for anything to challenge the environment in importance in any ranking of primacy. Therefore, the status quo should stay—ESG!

However, GES or ESG, the importance of corporate governance cannot be overemphasized. As a result, today's corporate champions must be able to strive not only for their pockets but for their planets too. They must be able to perceive both the financial and non-financial values of ESG.[11] To succeed, they must make sustainability an integral part of their business models, develop interest in as much of the community, country, world, and value chain as they can. This sounds like stakeholder capitalism and yes indeed. However, we must be careful because most stakeholders are neither investors nor risk takers in the companies they are said or they say they "hold stakes" in. Not having a financial stake in a company in

[11] Prall argues that the current accounting framework must be changed to take both tangible assets (which have relatively high but capped rates of return and are risky) and intangible assets (which are scalable) through ESG), Kevin Prall, "Perspectives Paper: A Framework to Assess ESG Value Creation," International Valuation Standards Council (IVSC), May 2021. Somewhat related to this are: the call by Becchetti stressing the "S" pillar in ESG (Leonardo Becchetti et al., "Going Deeper into the S of ESG: A Relational Approach to the Definition of Social Responsibility," *Sustainability*, Vol. 14 (2022): 9668, https://doi.org/10.3390/su14159668; The World Business Council for Sustainable Development (WBCSD) & The Institute of Internal Auditors (IIA), "Embedding ESG and Sustainability Considerations into the Three Lines Model," *WBCSD*, July 2022.

terms of an investment that costs you money may provide benefits to you but its losses may not hurt you the same way as someone who has put his/her money in that company. The effects are different even among the financing stakeholders (investors) depending on the share structure and non-financing stakeholders such as employees, suppliers, local government officials, and a neighbouring community depending on how much benefit they derive from the activities of a company. As a result, forcing the real investors to treat the non-financing stakeholders like investors or risk takers in the financial sense can affect entrepreneurship and investments incentives negatively. What we need is to ensure that the real investors and risk takers (who sometimes go through extreme challenges to execute projects) feel an adequate sense of compensation from their struggles, whilst not forgetting others are out there.

A win–win game is what is needed, but the players need to know their relative contributions and agree on their expectations. Naturally, it is always easier for a passenger than a driver to criticize how a car looks like and how it is driven. Win–win investment outcomes for stakeholders, based on relative contributions, is not easy but possible for all the reasons we put forward in this book and for the fact that where there is the will there should be the way.

8.2.9 The Cost of Carbon Dioxide Capitalism

Next, is about carbon. Carbon credits are a new means of fighting global climate change through economic incentives. They are a way of encouraging businesses and other institutions to conserve the environment and limit the impact of their activities on it through self-restraint as a result of costs (for companies) or benefits (other organizations) through the carbon credit system. Essentially, carbon credits focus on companies seeking to become carbon neutral, climate positive, climate neutral or net-zero, the main difference between which is whether the focus is on carbon emissions only or greenhouse gas (GHG) emissions as a whole into the atmosphere. The problem with carbon credit schemes is that the carbon credit has to be priced in manners that would truly discourage companies from emissions. If companies find it much better or easier to pay for their emissions cheaply, some may use carbon finance simply for marketing purposes.

Therefore, to work for the benefit of the environment and earth, a company must be involved not only in a carbon credit scheme but in other direct and internal reduction schemes to enable it to go beyond only paying for its pollution; in other words, attain a negative impact on emissions and positive position in relation to the environment.

8.2.10 *The Paradox of Technology*

Tenthly, the most distinguished and distinguishing feature of modern economic development is technology. Human civilizations in the past have advanced in many ways but never beyond the levels modern technology has reached, except perhaps in a very few areas. Yet, as we brag and boast about our technology, we tend to focus on the positive sides in our celebrations and forget the negative consequences or downsides. That is because we also forget how we have invented, developed, advanced, and utilized the technology to use, abuse, misuse, and overuse the abundant natural resources on earth, on which our lives depend but which we have inherited as humanity at no cost. In short, modern technology has enabled us to perform miracles by boosting our abilities to do things and reach frontiers that we have never imagined. Our mentalities, academic and working environments, markets, and economies in general have all been conducive to the incredible growth in the pace of technological advancements in the last 50 or so years. This has yielded incredible benefits to humankind, but the truth is that we have also used the technology to engage in unnecessary and destructive lifestyles and activities ranging from war to waste. We have (and continue to) used technology to bully and oppress each other and to exploit and prove that we can exploit natural resources anywhere, anyhow, and anyway we want to. This is at all levels, from investors, managers, regulators, to consumers. Without the necessary technology, this would never have been possible because numerous civilizations ended in history for many reasons but none have done so by extinguishing itself because of the paradox of its means of production; i.e. being able to develop and deliberately use its technology to destroy itself. The main reason why this is happening to us is because technology chases capital most of the time instead of vice versa. Here comes another paradox, whereas technology dominates our lives today because we have reached a stage where we cannot live without its essentials and new pleasures, necessities, and conveniences, it is dominated by capital, which is dominated and directed by not too many of us. So, to free the environment from the oppression of technology, technology must be freed from capital and capital must be freed from the dominance of

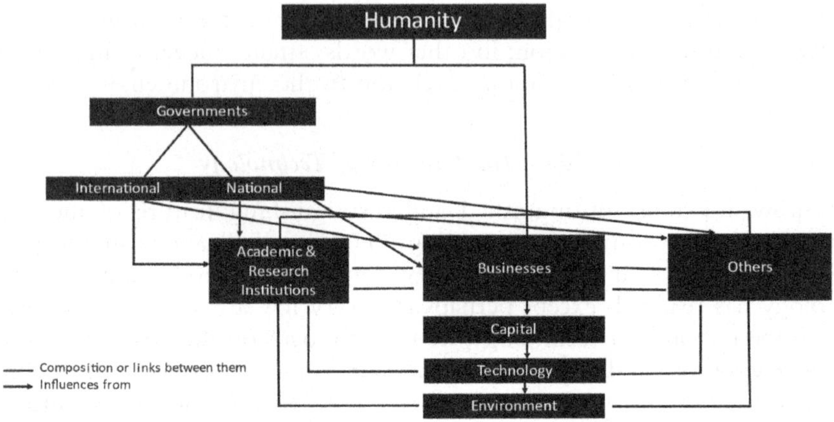

Fig. 8.3 Institutions, capital, technology and the environment (*Source* Author)

the profit motive when it comes to the commons of humanity. In that way, technology can move freely where it matters the most in tackling the urgent challenges/problems of climate change and other issues of the world.

Therefore, we suggest that the main solutions to these paradoxes be sought from governments, wealthy financiers, bright brains, and good managers, the sources of the contributions to the evolution of modern technology in the last 50 or so years. These solutions must be based on the answers to the question: "How can we make modern technology serve humankind and the environment much better?" The answers, like the technology itself, lie in pockets, heads, and hands where the technology originates and where it is directed from (Fig. 8.3).

8.2.11 The Paradox of Poverty

Eleventh, there is another paradox—the paradox of poverty in the global climate debate. By the paradox of poverty, we are referring to the situation of poor countries, particularly in Africa, who are the least developed, least polluting, but worst victims of global climate change, as we have pointed out in Chapter 2. At the same time, the need and urgency to develop these countries are greatest for them than the other/more advanced countries. As the African Union (2022) itself expressed it, "African cities, while currently responsible for a negligible amount of total global GHG emissions, are also under significant threat from climate change, which significantly exacerbates all the conditions …While Africa has achieved significant development gains in recent decades, the continent still lags behind other world regions in terms of many social and economic development indicators".[12] Africa is calling for huge amounts of money to cope with the effects of climate change, about $300 for only 35 Ethiopian, Kenyan, and South African cities by 2050 and it is asking for the international community (developed countries mainly) to pay the bills as they are the ones polluting the world the most. Africa is accusing the latter of not doing enough, very far from enough. African countries also accuse them for not doing enough in the actual reduction of GHG emissions from their sides, so the finger-pointing is in all directions and there is a *quid pro quo* attitude towards each other on unclear terms.

We believe that it is logical and sensible that those who pollute the most, cut the most, and pay the most in the reduction and clean up. However, as we declare next, everyone has a role to play.

8.2.12 The Verdict: We Are All Guilty

Finally, guilt, based on experience again. It has happened to us many times in a restaurant! Before we finish eating our plates are snatched from the table. Several times when we drew attention to the fact that we were still eating, we were politely told to go and take clean plates. In hotel rooms, bedsheets, blankets, towels, and soap are changed daily unnecessarily. In our offices workers live the lights on the whole night, close, walk out, and back into them the next day without even noticing it. At home adults who got used to sleeping with lights on forget or leave them on when they go to bed as adults, water is wasted in kitchens, showers, and bathtubs without the slightest hesitation. Across wherever it can be afforded, food is wasted, whereas wherever it cannot be afforded (which is much of the world), people are craving for it if not yes, the biggest companies and other establishments pollute the most, but largely in support of the lifestyles that we crave for like the missing grains in the desert where

[12] AU Climate Change and Resilient Development Strategy and Action Plan 2022–2032, https://www.tralac.org/documents/resources/african-union/4566-au-climate-change-and-resilient-development-strategy, retrieved April 5, 2023.

they cannot be grown or afforded. Yes, individual and national contributions and contradictions are different from country and continent to another,[13] but we are all or almost all contributing (or perhaps would all if we were given the opportunities to contribute) to the challenges of the world, including those in Africa, that ESG seeks to address. We have seen in some of the most isolated villages and poorest suburbs of the world, how individuals living from hand to mouth own or seek or own the expensive phones, television sets, clothing, cars, etc. Our aims and aspirations have become largely universalized in an interdependent and interconnected global village. Everyone (except perhaps those who eke out their living from the earth, subsisting on livestock, game meat, or fish without schools, televisions, laptops, and mobile phones) have a role to play for a better all for all, especially in the area of the environment. You may not see it but your five toes are in that carbon footprint you may be attributing only to others, most likely it is. Yes, we are all (or almost all) indeed guilty!

Therefore, to reiterate, all of us have a role to play in dealing with the ESG challenges, especially environmental, of this world. However, the biggest challenge we may all be having in carrying our responsibilities is that when a man walks more than half the wrong way to a destination, it becomes very difficult to bring him back to look for another route. He would rather keep on going, because of the convenience and hope that his distance would be closer going the same way. We have experienced this a number of times as village kids when walking on footpaths looking for a village and getting lost on the way! You keep on hoping and going! That is what is happening in our world today; we are hoping, going, and even expecting that businesses would be the ones to rescue all of us through ESG. Yes, companies and other organizations have very big roles through ESG, particularly in Africa, and that is why we write this book, but almost every individual in the continent and the rest of the world has a role too at the personal level.

[13] Indeed, according to the International Resource Panel, set up by the United Nations Environment Programme (UNEP), citizens of developed countries consume an average of 16 tonnes of minerals, ores, fossil fuels and biomass per annum per capita per annum. This is more than 40 tonnes per person in some of the countries. On the other hand, the average Indian today consumes four tonnes per year. UNEP, *Decoupling Natural Resource Use and Environmental Impacts from Economic Growth. A Report of the Working Group on Decoupling to the International Resource Panel*, 2011.

Bibliography for Additional Reading

African Development Bank (AfDB). (2018a). *Report: African Nationally Determined Contributions (NDCs)*. Abidjan: AfDB.

African Development Bank (AfDB). (2018b). *Roadmap and Work Programme: Africa NDC Hub, AfDB*. Abidjan: AfDB.

African Development Bank (AfDB). (2000). *Bank Group Policy on Good Governance*. Abidjan: AfDB.

African Development Bank (AfDB). *African Economic Outlook 2022: Supporting Climate Resilience and a Just Energy Transition in Africa*. Abidjan: AfDB.

African Development Bank (AfDB). *African Economic Outlook 2023: Mobilizing Private Sector Financing for Climate and Green Growth in Africa*. Abidjan: AfDB.

African Union Commission. (2015). *Agenda 2063: The African We Want: First Ten-Year Implementation Plan, 2014–2023*. Addis Ababa: African Union Commission.

Aguilar, Francis J. (1967). *Scanning the Business Environment*. New York, NY: Macmillan.

Al-Tuwaijri, S. A., T. E. Christensen, & K. E. Hughes. (2004). "The Relations Among Environmental Disclosure, Environmental Performance, and Economic Performance: A Simultaneous Equations Approach." *Accounting, Organizations and Society*, vol. 29, no. 5–6, pp. 447–471.

Alexander, Gordon J., & Rogene A. Buchholz. (1978). "Corporate Social Responsibility and Stock Market Performance." *The Academy of Management Journal*, vol. 21, no. 3, pp. 479–486.

K. N. M. Sonko and M. Sonko, *Demystifying Environmental, Social and Governance (ESG)*, Palgrave Studies in Impact Finance, https://doi.org/10.1007/978-3-031-35867-8

Barnett, M. L., & H. Salomon. (2002). "Beyond the Dichotomy: The Curvilinear Relationship Between Social Responsibility and Financial Performance." *Strategic Management Journal*, vol. 27, pp. 1101–1122.

Bassen, A., & M. Kovàcs. (2008). "Environmental, Social and Governance Key Performance Indicators from a Capital Market Perspective." *ZeitschriftfürWirtschafts-und Unternehmensethik*, vol. 9, no. 2, pp. 182–192.

Boateng, Frank. (2023). "Leveraging the 'G' in ESG for Growth and Sustainability in Africa." Presentation at the 2nd High-Level Webinar on "ESG and Stakeholder Engagement in Africa," May 19, 2023.

Bowman, E. H., & M. Haire. (1975). "A Strategic Posture Toward Corporate Social Responsibility." *California Management Review*, vol. 18, no. 2, pp. 49–58.

Bradley, Brendan. (2021). *ESG Investing For Dummies (For Dummies (Business & Personal Finance))*. Hoboken, NJ: For Dummies.

Bryman, A., & E. Bell. (2011). *Business Research Methods*, Ed. 3. New York, NY: Oxford University Press.

Câmara, Paulo, & Filipe Morais, eds. (2022). *The Palgrave Handbook of ESG and Corporate Governance*. New York, NY: Palgrave Macmillan.

Carroll, A. B. (1979). "Three Dimensional Conceptual Model of Corporate Performance." *Academy of Management Journal*, vol. 4, no. 4, pp. 497–505.

Chang, Ha Joon. (2023). *Edible Economics: A Hungry Economist Explains the World*. New York, NY: PublicAffairs.

Chung, Kee H., & Stephen W. Pruitt. (1994). "A Simple Approximation of Tobin's Q." *Financial Management*, vol. 23, no. 3, pp. 70–74.

Clay, R. (1994). "A Continent in Chaos: Africa's Environmental Issues." *National Library of Medicine*, December 1994. https://pubmed.ncbi.nlm.nih.gov/7713013/.

Core, J. E., Guay, W. R., & Rusticus, T. O. (2006). "Does Weak Governance Cause Weak Stock Returns? An Examination of Firm Operating Performance and Investor's Expectations." *The Journal of Finance*, vol. 61, no. 2, pp. 655–687.

Daggash, Habiba Ahut, and Niall Mac Dowell. (2021). "Delivering Low-Carbon Electricity Systems in Sub-Saharan Africa: Insights from Nigeria." *Energy & Environmental Science*, vol. 14, pp. 4018–4037.

Derwall, J. (2007). *The Economic Virtues of SRI and CSR, Erasmus University Rotterdam*. Erasmus Research Institute of Management. Available at: http://repub.eur.nl/pub/8986. Accessed April 3, 2014.

Devereux, Stephen. (2000). *Social Safety Nets for Poverty Alleviation in Southern Africa: A research report for DFID, ESCOR Report R7017*. Brighton, Sussex: The Institute of Development Studies at the University of Sussex.

Devereux, Stephen. (2011). "Social Protection in South Africa: Exceptional or Exceptionalism?" *Canadian Journal of Development Studies/Revue canadienne d'études du développement*, vol. 32, no. 4, pp. 414–425.

Ding, David K., Christo Ferreira, & Udomsak Wongchoti. (2016). "Does It Pay to Be Different? Relative CSR and Its Impact on Firm Value." *International Review of Financial Analysis*, vol. 47, pp. 86–96.

Dubai Electricity and Water Authority (DEWA). *Sustainability Reports 2020 and 2021*, Dubai 2020. Dubai: DEWA.

Dubai Government (UAE). *Expo2020 Dubai Sustainability Report*, Dubai 2020. Dubai: DEWA.

Eccles, R. G., M. P. Krzus, J. Rogers, & G. Serafeim. (2012). "The Need for Sector-Specific Materiality and Sustainability Reporting Standards." *Journal of Applied Corporate Finance*, vol. 24, no. 2, pp. 65–71.

Elkington, John. (1998). *Cannibals with Forks: The Triple Bottom Line of the 21st Century Business*. Gabriola, BC: New Society Publishers.

Fields, Scott. "Continental Divide: Why Africa's Climate Burden Is Greater." *Environmental Health* Perspectives, vol. 113, no. 8. https://www.ncbi.nlm.nih.gov/pmc/articles/PMC1280367/.

Flammer, C. (2013). "Corporate Social Responsibility and Shareholder Reaction: The Environmental Awareness of Investors." *Academy of Management Journal*, vol. 56, no. 3, pp. 758–781.

Freeman, R. E. (1984). *Strategic Management: A Stakeholder Approach*. Boston, MA: Pitman.

Freeman, R., & J. McVea. (2001). "A Stakeholder Approach to Strategic Management." Working Paper No. 01-02. Darden Graduate School of Business Administration, University of Virginia, Charlottesville, VA.

Friedman, M. (1970). "The Social Responsibility of Business Is to Increase its Profits." *The New York Times Magazine*, September 13.

Fry, Terry/Cadmus. (2023). "Increasing Energy Access and Energy Security." Presentation at the 2nd High-level Webinar on "ESG and Stakeholder Engagement in Africa," May 19, 2023.

Galbreath, J. (2013). "ESG in Focus: The Australian Evidence." *Journal of Business Ethics*, vol. 118, pp. 529–541.

Garth Meyers. (2019). "Environmental Issues in Africa's Cities." *Wiley Online Library*, April 2019. https://doi.org/10.1002/9781118568446.eurs0002.

Gisselquist, Rachel. (2012). "What Does 'Good Governance' Mean?" United Nations University, January 2012.

Globe Afrique. (2016). "Africa Faces Some Serious Environmental Problems: Greenpeace." *Globe Afrique*, November 2016. https://globeafrique.com/africa-faces-some-serious-environmental-problems-greenpeace/.

Gompers, P., Ishii, J., & Metrick, A. (2003). "Corporate Governance and Equity Prices." *The Quarterly Journal of Economic*, February, pp. 107–155.

Grant, Richard. (2015). *Africa: Geographies of Change*. New York, NY: Oxford University Press.

Graves, Samuel, & Sandra A. Waddock. (1994). "Institutional Owners and Corporate Social Performance." *The Academy of Management Journal*, vol. 37, no. 4 (Aug., 1994), pp. 1034–1046.

Guenster, Nadja, Rob Bauer, Jeroen Derwall, & Kees Koedijk. (2011). "The Economic Value of Corporate Eco-Efficiency." *European Financial Management*, vol. 17, no. 4, pp. 679–704.

Haque, Faizul. (2017). "The Effects of Board Characteristics and Sustainable Compensation Policy on Carbon Performance of UK Firms." *The British Accounting Review* 49, pp. 347–364.

Hillman, A. J., & G. D. Keim. (2001). "Shareholder Value, Stakeholder Management and Social Issues: What's the Bottom Line?" *Strategic Management Journal*, vol. 22, no. 2 (February), pp. 125–139.

Holder-Webb, L., J. Cohen, L. Nath, & D. Wood. (2009). "The Supply of Corporate Social Responsibility Disclosures among U.S. Firms." *Journal of Business Ethics*, vol. 84, no. 4, pp. 497–527.

International Labour Organization (ILO). (2017). *World Social Protection Report 2017–19: Universal Social Protection to Achieve the Sustainable Development Goals*. Geneva: ILO.

ILO. (2019). *Extending Social Security to Workers in the Informal Economy Lessons from International Experience*. Geneva: ILO.

ILO. (2021). *Africa Regional Social Protection Strategy, 2021–2025: Towards 40 %—A Social Protection Coverage Acceleration Framework to Achieve the SDGs*. Geneva: ILO.

ILO. (2021). *World Social Protection Report 2020–22: Social Protection at the Crossroads—In Pursuit of a Better Future*. Geneva: ILO.

International and Institute of Development Studies (ODI), Save the Children UK, and HelpAge. (2005). "*Making Cash Count: Lessons from Cash Transfer Schemes in East and Southern Africa for Supporting the Most Vulnerable Children and Households*." Institute of Development Studies, University of Sussex, Brighton.

International Renewal Energy Agency (IRENA) and the African Development Bank (AfDB). (2022). *Renewable Energy Market Analysis: Africa and its Regions*. Abu Dhabi: IRENA.

Kabir Hassan, M., Aishath Muneeza, & Karamo N. M. Sonko. (2022). *Islamic Finance in Africa: The Prospects for Sustainable Development*. Edward Elgar Publishing.

Kende-Robb, Caroline. (2015). "Five Steps to Save Africa from Climate Change." World Economic Forum, December 2015. https://www.weforum.org/agenda/2015/12/5-steps-to-save-africa-from-climate-change/.

Khan, H. A., & Karamo N. M. Sonko. (1994). "A Further Extension of Adjustment Models: The Environment and Equity." In *Economic Justice in Africa: Adjustment and Sustainable Development*. Westport, CT: Greenwood Press.

Khan, H. A., & Karamo N. M. Sonko, eds. (2021). *Islamic Finance as a Complex System: New Insights*. Lanham, MD: Lexington Books.

Khan, Haider A., & M. Parvin. (1990). "Environment, Externalities and Income." Unpublished paper, University of Denver, CO.

Kocmanova, A., & I. Simberova. (2012). "Modelling of Corporate Governance Indicators." *Engineering Economics*, vol. 23, no. 5, pp. 485–495.

KPMG. (2011). "Sustainable Insight: Articulating the Value of Sustainability to Mainstream Investors." Available at: https://www.kpmg.com/GR/en/IssuesAndInsights/Articles,Publications/Sustainability/Documents/Sustainable-Insight_Jan_2011.pdf. Accessed April 24, 2014.

La Porta, R., F. Lopez-De-Silanes, A. Shleifer, & R. Vishny. (1998). "Law and Finance." *Journal of Political Economy*, vol. 106, no. 6, pp. 1113–1155.

La Porta, R., F. Lopez-De-Silanes, A. Shleifer, & R. Vishny. (2002). "Investor Protection and Corporate Valuation." *The Journal of Finance*, vol. 57, no. 3, pp. 1147–1169.

Labuschagne, Carin, Alan Brent, & Ron P. G. van Erck. (2005). "Assessing the Sustainability Performances of Industries." *Journal of Cleaner Production*, vol. 13, no. 4, pp. 373–385.

Lamere, Carolyn. (2013). "UNEP Highlights Environmental Impacts on Health in Africa." New Security Beat, March 2013. https://www.newsecuritybeat.org/2013/03/unep-highlights-environmental-impacts-health-africa/.

Locke, Natania. (2020). "Corporate Governance in South Africa." European Corporate Governance Institute (ECGI), September 2020. https://ecgi.global/content/corporate-governance-south-africa#:~:text=Corporate%20governance%20in%20South%20Africa%20is%20informed%20by%20common%20law,in%20financial%20and%20market%20regulation. Last visited January 15, 2023.

London Stock Exchange Group (LSEG). (2020). *Revealing the Full Picture. Your Guide to ESG Reporting Guidance for Issuers on the Integration of ESG into Investor Reporting and Communication*. London.

Lopez-de-Silanes, Florencio, Andrei Shleifer, Rafael la porta, & Robert W. Vishny. (2000). "Agency Problems and Dividend Policies Around the World." *The Journal of Finance*, vol. 55, no. 1, pp. 1–33.

Mabogunje, Akin. (1998). Environmental Challenges in Sub-Saharan Africa." Africa Technology Forum, May 1998. http://web.mit.edu/africantech/www/articles/EnvChall.htm.

Margolis, J. D, & J. P. Walsh. (2003). "Misery Loves Companies: Rethinking the Social Initiatives by Business." *Administrative Science Quarterly*, vol. 48, pp. 268–305.

Matos, Pedro Matos. (2020). "ESG and Responsible Institutional Investing Around the World: A Critical Review." CFA Institute Research Foundation.

Mbaka, Cecilia, et al. (2017). *Kenya Social Protection Sector Review: 2017*. Edited by John Gachigi, 2018—Development Pathways, Republic of Kenya Ministry of Labour and Social Protection State Department for Social Protection.

McWilliams, A., & D. S. Siegel. (2000). "Corporate Social Responsibility and Finance Performance: Correlation or Misspecification?" *Strategic Management Journal*, vol. 21, pp. 603–609.

Mo Ibrahim Foundation. (2018). *Ibrahim Index of African Governance 2018*. London: Mo Ibrahim Foundation.

Motlanthe, Kgalema. (2022). "Remarks at the First High-level Webinar on ESG and Stakeholder Capitalism in Africa." Virtual Conference organized by Heeno International, Jula Consultancy FZE and partners on January 28, 2022.

Mutiso, Rose, Ken Caldeira, Murefu Barasa, Moussa Blimpo, Lauren Culver, Habiba Ahut Daggash, Michael Dioha, Zeke Hausfather, June Lukuyu, Joel Nana, Joan Nkiriki, Lily Odarno, & Katie Auth. (2022). "Who Decides Africa's Net Zero Pathways? Five Ways to Fix How We Model African Energy Transitions and Why It Matters for Climate and Development." Washington, DC: Energy for Growth Hub.

Nelling, E., & E. Webb. (2009). "Corporate Social Responsibility and Financial Performance: The "Virtuous Circle" Revisited." *Review of Quantitative Finance and Accounting*, vol. 32, pp. 197–209.

OECD. (2001). "Policies to Enhance Sustainable Development." Organization for Economic Co-Operation and Development, April 2001. https://www.oecd.org/greengrowth/1869800.pdf.

Okonjo-Iweala, Ngozi. (2020). "Africa Can Play a Leading Role in the Fight against Climate Change." Brookings Institution. https://www.brookings.edu/research/africa-can-play-a-leading-role-in-the-fight-against-climate-change/.

Orlitzky, Marc, & John D. Benjamin. (2001). "Corporate Social Performance and Firm Risk: A Meta-Analytic Review." *Business & Society*, vol. 40, no. 4, pp. 369–396.

Orlitzky, M., F. Schmidt, & S. L. Rynes. (2003). "Corporate Social and Financial Performance: A Meta-analysis." *Organization Studies*, vol. 24, no. 3 (March), pp. 403–441.

Ortiz, Isabel. (2018). "The Case for Universal Social Protection: Everyone Faces Vulnerabilities During Their Lifetime." *Finance & Development*, vol. 55, no. 4, pp. 32–34.

Patel, Leila. (2019). "Lessons from South Africa on Why Gender Matters in Social Welfare Policies." *The Conversation*, October 21, 2019.

Pettingale, Stéphane de Maupeou, & Peter Reilly. (2019). "EU Taxonomy and the Future of Reporting." Harvard Law School Forum on Corporate Governance, April 4, 2022.

Purvis, Ben, Yong Mao, & Darren Robinson. (2019). "Three Pillars of Sustainability: In Search of Conceptual Origins." *Sustainability Science*, vol. 14, 681–695.

Renneboog, L., J. Ter Horst, & C. Zhang. (2008a). "Socially Responsible Investments: Institutional Aspects, Performance and Investor Behavior." *Journal of Banking and Finance*, vol. 32, pp. 1723–1742.

Renneboog, L., J. Ter Horst, & C. Zhang. (2008b). "The Price of Ethics and Stakeholder Governance: The Performance of Socially Responsible Mutual Funds." *Journal of Corporate Finance*, vol. 14, pp. 302–322.

Roberts, Peter W., & Grahame R. Dowling. (2002). "Corporate Reputation and Sustained Superior Financial Performance." *Strategic Management Journal*, vol. 23, no. 12, pp. 1077–1093.

Roser, M., & H. Ritchie. (2018). "HIV/AIDS." *Our World in Data*, revised in November 2019. https://ourworldindata.org/hiv-aids. Accessed June 7, 2023.

Savitz, Andrew, and Karl Weber. (2013). *The Triple Bottom Line (revised and updated)*. Hoboken, NJ: Jossey-Bass.

Schwab, Klaus, & Peter Vanham. (2021). *Stakeholder Capitalism: A Global Economy that Works for Progress, People and Planet*. Hoboken, NJ: Wiley.

Seo, K., J. Moon, & S. Lee. (2015). "Synergy of Corporate Social Responsibility and Service Quality for Airlines: The Moderating Role of Carrier Type." *Journal of Air Transport Management*, vol. 47, pp. 126–134.

Sonko, Karamo N. M. (2020a). "Next Energy Capital: Facing the Challenges with PESTLE & Mortar." London, presentation on March 10, 2020.

Sonko, Karamo N. M. (2020b). "The African PESTLE for Meyas Sand Minerals Company Ltd." Presentation to Senior Managament Team, Khartoum, Sudan, February 6, 2020.

Sonko, Karamo N. M., & Mariama Sonko. (2016). "BADEA in a New African Economy: Facing the Challenges with PESTLE & Mortar." Presentation to BADEA, May 311, 2016; and Mariama Sonko, "ESG: Relevance and Applicability in Africa," Presentation to BADEA, July 5, 2021.

Sonko, Karamo N. M., & Mariama Sonko. (2019). "Al Dhabi Investments Ltd: Facing the Management Challenges with Arabian PESTLE & Mortar," December 8, 2019.

Sonko, Karamo N. M., & Mariama Sonko. (2019). "Samil Industrial Co. Ltd: Facing the Challenges with African PESTLE & Mortar," December 26, 2019.

Sonko, Karamo N. M., & Mariama Sonko. (2020a). "Strategic Management at Corporate & Personal Levels: Facing the Challenges with the African

PESTLE & Mortar." Online Presentation to FUFC, Abu Dhabi, Wednesday, July 27, 2020.

Sonko, Karamo N. M., & Mariama Sonko. (2020b). "Strategic Management in Your Company or Organization: Facing the Challenges with the African PESTLE Analysis." Pilot Session, Etisalat Academy, December 30, 2020.

Sonko, Karamo N. M., & Mariama Sonko. (2021a). "Strategic Management in Your Company or Organization: Facing the Challenges with the African PESTLE Analysis." Presentation to UNIDO AICEI, Bahrain, May 31, 2021.

Sonko, Karamo N. M., & Mariama Sonko. (2021b). "Strategic Management: African PESTLE Analysis for African Governments & Regional Organizations." Presentation to ACBF, Harare, Zimbabwe, March 19, 2021.

Sonko, Karamo N. M., & Mariama Sonko. (2021c). "The African PESTLE Analysis." Online Presentation to the OACPS, Brussels, Belgium, March 26, 2021.

Sonko, Karamo N. M., & Mariama Sonko. (2021d). "Doing Business in Africa: Opportunities, Risks and Strategies." Presentation to the Abu Dhabi Islamic Bank, Khartoum, Sudan, October 11, 2021.

Sonko, Karamo N. M., & Mariama Sonko. (2021e). "The African PESTLE Analysis." Presentation to GRAS, Ministry of Minerals, Khartoum, Sudan, May 25, 2021.

Sonko, Karamo N. M., & Mariama Sonko. (2021f). "Promoting Domestic and Foreign Investment in the Sudanese Mining Sector." Presentation to SMRC, Ministry of Minerals, Khartoum, Sudan, 11 August 2021.

Sonko, Mariama. (2021a). "ESG Disclosure, ESG Performance and Firm Value: Evidence from South Africa." M.Sc. dissertation, Department of Finance, Heriot-Watt University, Dubai, 2018.

Sonko, Mariama. (2021b), "ESG: Relevance and Applicability in Africa." Arab Bank for the Economic Development of Africa (BADEA), Khartoum, Sudan, July 10, 2021.

Sy, Amadou. (2016). "Africa: Financing Adaptation and Mitigation in the World's Most Vulnerable Region." Brookings Institution, February 2016. https://www.brookings.edu/wp-content/uploads/2016/08/global_20160818_cop21_africa.pdf.

Turban, D., & D. W. Greening. (1996). "Corporate Social Performance and Organizational Attractiveness to Prospective Employees." *Academy of Management Journal*, vol. 40, pp. 658–672.

Ullman, A. H. (1985). "Data in Search of a Theory: A Critical Examination of the Relationship among Social Performance, Social Disclosure, and Economic Performance of US Firms." *Academy of Management Review*, vol. 10, no. 3, 1985, pp. 540–557.

UNEP. (2020). "Africa: The Challenge." United National Environment Program. https://www.unenvironment.org/regions/africa#:~:text=Africa%20faces%20serious%20environmental%20challenges,extreme%20vulnerability%20to%20climate%20change.

United Nations. (2018). *Promoting Inclusion Through Social Protection: Report on the World Social Situation 2018*. New York, NY: UN.

United Nations Development Program (UNDP). (2018). *Human Development Index 2018*. New York: Oxford University Press.

UNU-WIDER. (2016). *Growth and Poverty in Sub-Saharan Africa*. Oxford: Oxford University Press.

Waddock, S., & S. B. Graves. (1997). "The Corporate Social Performance—Financial Performance Link." *Strategic Management Journal*, vol. 18, no. 4, pp. 303–319.

Werbach, Adam. (2009). *Strategy for Sustainability: A Business Manifesto*. Cambridge, MA: Harvard University Press.

Werbach, Adam. (2010). "What Is a Sustainable Brand?" Presentation from the National Green Brands Forum, Melbourne, June 17, 2010.

World Bank. (2019a). "Eliminating Gender Disparities in Business Performance in Africa: Supporting Women-Owned Firms." World Bank. https://www.worldbank.org/en/region/afr/publication/eliminating-gender-disparities-in-business-performance-in-africa-supporting-women-owned-firms.

World Bank. (2019b). "Accelerating Poverty Reduction in Africa: In Five Charts." World Bank, October 9, 2019. https://www.worldbank.org/en/region/afr/publication/accelerating-poverty-reduction-in-africa-in-five-charts.

World Bank. (2023). World Bank Group Strategy for Fragility, Conflict, and Violence 2020–2025. https://www.worldbank.org/en/topic/fragilityconflictviolence/overview.

Yang, Ann Shawing, & Suvd Baasandori. (2017). "Exploring CSR and Financial Performance of Full-Service and Low-Cost Air Carriers." *Finance Research Letters*, vol. 23, pp. 291–299.

Yoon, Bohyun, Jeong Hwan Lee, & Ryan Byun. (2018). "Does ESG Performance Enhance Firm Value? Evidence from Korea." *Sustainability*, vol. 10, no. 10, pp. 1–18.

UNEP [illegible] The Challenge [illegible]
[illegible]
United Nations [illegible]
[illegible]
World Bank [illegible]
[illegible]

Index

K. N. M. Sonko and M. Sonko, *Demystifying Environmental, Social and Governance (ESG)*, Palgrave Studies in Impact Finance,
https://doi.org/10.1007/978-3-031-35867-8

The manufacturer's authorised representative in the EU is Springer Nature Customer Service Centre GmbH, Europaplatz 3, 69115 Heidelberg, Germany. If you have any concerns regarding our products, please contact ProductSafety@springernature.com

Printed and bound by CPI Group (UK) Ltd, Croydon, CR0 4YY
28/09/2026
02224074-0006